THE SPIRIT OF THOREAU

Material Faith

The Spirit of Thoreau

SPONSORED BY THE THOREAU SOCIETY

Wesley T. Mott, Series Editor

Uncommon Learning:
Thoreau on Education

Material Faith:
Thoreau on Science

Elevating Ourselves:
Thoreau on Mountains

Material Faith

THOREAU ON SCIENCE

Edited by Laura Dassow Walls

Foreword by Edward O. Wilson

A Mariner Original

Houghton Mifflin Company

BOSTON   NEW YORK

1999

For information about permission to reproduce selections
from this book, write to Permissions, Houghton Mifflin Company,
215 Park Avenue South, New York, New York 10003.

Library of Congress Cataloging-in-Publication Data

Thoreau, Henry David, 1817–1862.
Material faith : Thoreau on science / edited by
Laura Dassow Walls ; foreword by Edward O. Wilson.
p.   cm. — (The spirit of Thoreau)
"A Mariner Original."
ISBN 0-395-94800-2
1. Thoreau, Henry David, 1817–1862 — Quotations.
2. Science — Quotations, maxims, etc.   3. Quotations, American.
I. Walls, Laura Dassow.   II. Title.
III. Series: Thoreau, Henry David, 1817–1862. Spirit of Thoreau.
PS3042.W34   1999   818'.309 — dc21   99-12267   CIP

Book design by Anne Chalmers
Type: Bulmer (Monotype)
The ferns and twigs on the title page
were gathered at Walden Pond.

Printed in the United States of America
QBP 10 9 8 7 6 5 4 3 2 1

Selections from volumes 1–5 of Thoreau's *Journal*
are from *The Writings of Henry David Thoreau*, Witherell, E., ed.
Copyright © 1972 by Princeton University Press. Reprinted
by permission of Princeton University Press.

Edward O. Wilson

Henry David Thoreau, whose spirit this series of books celebrates, was thought by many in his own time to be an eccentric who escaped from the mainstream of real life in order to dream. He was the opposite of that. He understood intuitively what we now know in more concrete and objective terms, that humanity is a biological species and thus exquisitely adapted to the natural world that cradled us. Thoreau was the scientific observer and lyrical expositor who hit upon the power of this conjunction between science and the humanities. He was the first great nature writer, whose knowledge of the living world, based on experience, was refined and projected as poetry. Nature writing, one of the major innovations of American literature, also includes in its pantheon John Muir, Aldo Leopold, and Rachel Carson. Together these writers say to us that humanity coevolved with the rest of life on this particular planet; other worlds are not in our genes. It is a delusion that people can flourish apart from the living world. We might do so physically, like animals in a feed lot, but not spiritually, not to the full extent for which our brains are designed.

People travel into nature in search of new life and wonder, and from nature they return to the parts of the earth that have been humanized and made physically secure. Nature, and especially that part saved as wilderness, settles peace on the soul because it needs no help; it is beyond human contrivance. It is also a metaphor of unlimited opportunity, rising from the tribal memory of a time in which humanity spread across the world, valley to valley, island to island, godstruck, firm in the belief that virgin land went on forever beyond the horizon. That is very much an American dream, and one we will be wise to keep alive by the preservation of our wild heritage.

INTRODUCTION

The Man Most Alive

LAURA DASSOW WALLS

The earliest pages of Thoreau's *Journal* trace preoccu-
pations that would trouble and inspire him to the end of
his life: How do we transform facts into knowledge,
truth, even wisdom? How can we bridge the "chasm" between knowledge and ignorance? What is the nature of
reality? And where is the "bottom that will hold an anchor, that it may not drag"? (13 August 1838). These are
the classic questions of metaphysics: What is a world?
What elements and beings compose it?

Thoreau's mentor, Ralph Waldo Emerson, regarded
theories and ideas as the true reality, and the physical
world as evanescent and fleeting; but even as a student
Thoreau was skeptical of the "fair theories" of men. He
put his faith instead in the physical reality of nature, of
what he once called "terrene, titanic matter" (30 August
1856), and more often called "the wild." Both Thoreau's
skepticism and his faith inclined him toward science,
even as science was taking over from religion the great
questions of existence. His writings, from beginning to
end, pursue a continuing debate over the hopes and lim-
itations of science, the expression both of human arro-

gance over all creation and of love for earth and all its creatures.

The sweeping advances of science in Thoreau's day opened up a vastly expanded universe even while they hinted that the universe might soon be unified under the rule of one law, a law comprehensible to human reason. Geology was deepening time from thousands to millions of years; astronomy measured the distance to the nearest stars in light years; physics had united the mysterious forces of electricity and magnetism; voyages of exploration were uncovering a staggering variety of life forms, while biology was tracking the ultimate principles common to all life. Technologies such as the railroad and the telegraph were altering the face of the continent and the pace of human communication. Science was the talk of parlor and street corner, even as men of science were consolidating and elevating their pursuit into a profession reserved for the few who could afford university training, expensive texts and equipment, and attendance at professional meetings. At the very moment when Thoreau was investigating nature at Walden Pond, Louis Agassiz arrived from Europe and set about revolutionizing science in America, starting at Thoreau's alma mater, Harvard College. Thoreau himself became a recruit, trapping and shipping specimens from Walden Pond to Agassiz's assistant, James Elliot Cabot.

Science played a particularly important role in the United States, where government-funded exploring expeditions were charting the unknown lands of the West and scientists were assessing the natural resources of state after state, publishing a steady stream of reports on geology, plants, "noxious" insects, birds, mammals, and reptiles — surveys whose lack of depth often drew down Thoreau's contempt. Moreover, the nature of science itself was under debate. Was true science purely deductive, "top-down," and predictive, in the manner of mathematics, astronomy, and "natural philosophy," or physics? Or could true science also be inductive, "bottom-up," and descriptive, like natural history? The progress of science suggested that even the vagaries of natural history would someday be reducible to the operation of known laws.

Thoreau learned natural philosophy as a student at Harvard, but he was drawn to the border sciences of natural history: botany, zoology, entomology, ornithology, meteorology, physical geography. From the British geologist Charles Lyell, Thoreau learned ways of reading the past in the present, as well as ways of anticipating the future by watching how the "little strokes" dealt by nature accumulate over time. Thoreau also learned from Alexander von Humboldt and Charles Darwin, naturalists whose total immersion in nature allowed them to recognize patterns and relationships, producing a new kind of science that joined rather than separated man and nature, "subject" and "object." Thoreau was convinced that the perceiver could never be extracted from the object perceived, and he became a shrewd analyst of perception's subtle interplay. In his persistent attention to relational knowledge, Thoreau explored concepts such as plant communities, seed dispersal, and forest succession — novel concepts then, but fundamental to the science of ecology, which would emerge only at the end of the century.

Thoreau's earliest focus was on defining the *true* "man of science." The word *scientist,* coined in England in 1834, would not come into popular use until well after Thoreau's death in 1862; thus there was no single word for a person who specialized in the study of nature. Into this conceptual space Thoreau poured his intellectual ideals: as he wrote in his early essay "Natural History of Massachusetts," the true man of science would be brave and sympathetic, would possess "a deeper and finer experience," even an "Indian wisdom." Since true knowledge was learned not from books but by action in the world, no science could be understood apart from the biography of the scientist. In his *Journal* Thoreau asserted that "the sum of what the writer of whatever class has to report is simply some human experience, whether he be poet or philosopher or man of science. The man of most science is the man most alive, whose life is the greatest event" (6 May 1854).

This ideal set Thoreau at odds with the gentlemen of science who strove for an objectivity he believed was unattainable and undesirable: "It is ebb-tide with the scientific reports," he grumbled, "Professor —— in the chair" (5 March 1858). His likely target was Emerson's friend, the omnipresent Agassiz, who was often in Thoreau's thoughts as he struggled to understand the distribution of plants and animals. When he pondered the problem of how toad spawn could have found their way to a pool near the summit of Mt. Monadnock, he dismissed the great scientist's explanation: "Agassiz might say that they originated on the top" (3 June 1858). Thoreau's own material faith insisted upon a physical agency for such unlikely events, and two years later, confronting a similar problem at home in Concord, he welcomed Darwin's very different reasoning, that plants and animals spread themselves, even to the most remote places, by their own independent movements. Darwin's "development theory," Thoreau concluded approvingly, was more "philosophical," amounting as it did "to a sort of constant *new* creation" (18 October 1860). Thoreau preferred a "living earth" that was ever in the midst of creating itself.

Thoreau's positive ideal emerged early, to become a lifelong tenet of his faith: the poet would unite earth and sky, science and philosophy, generalizing "their widest deductions" and so crossing the "chasm between knowledge and ignorance" through his own practical experience and action in the world. The poet would turn science into "con-science," a *moral* knowledge. Paradoxically, it was this ideal that drew the poet Thoreau closer to the method of science, which sought to join fact and theory through the test of experience; "every poet," he mused, "has trembled on the verge of science" (18 July 1852). The turning point of his career as a poet-scientist came around 1850, when he became a scientific traveler within the boundaries of Concord, taking long daily walks, which he reenacted in his *Journal*. These walks were collecting expeditions: into his "botany box" (a rather disreputable straw hat) went all variety of plants. His attic room filled up with pressed plants; birds' nests and eggs; insects in various stages of hatching, growth, and preservation; turtle eggs and turtle shells; Indian artifacts; stuffed birds; and natural curiosities brought to him by his neighbors. On his occasional trips to Boston and Cambridge, Thoreau compared specimens in museums with those he had collected or seen and chatted with fellow members of the Boston Society of Natural History (to which he had been elected in 1850). His activities opened to his view a dizzying array of natural objects and phenomena—"How many questions there are which I have not put to the inhabitants!" (7 June 1851) he marveled.

Yet even as his *Journal* ripened, Thoreau recorded a sense of strain and loss: "I fear that the character of my knowledge is from year to year becoming more distinct & scientific," he worried (19 August 1851). Moments of anguish alternated with moments of elation: "I feel blessed. I love my life" (1 November 1852). Questions led to more questions, and every new fact suggested "that worlds remain to be unveiled" (19 April 1852). Since knowledge lay in experience rather than books, "each object appears wholly undescribed to our experience—each field of thought wholly unexplored— The whole world is an America—a *New World*" (2 April 1852). Against such a background, Thoreau transformed his lament "I am afraid to be so knowing" (20 August 1851) into an affirmative refrain: "I will not fear to *know*" (7 May 1852).

Thoreau in the early 1850s advanced confidently in the direction of science in the faith that the universe would bear his weight. There were disappointments: one day he discovered that some young pine shoots were all pointing to the east, a fact he considered of tremendous significance—but it was spoiled the next day when he found them all inclined to the west. Yet there were deep satisfactions, as when learning a new scientific name allowed him to "get hold" of some object for the first time. And he found a solution to the strain of overattention, too, in what he called "a true sauntering of the eye" (28 April 1856), by which the prepared mind might be relaxed enough to be caught off guard by the unforeseen. Thoreau's interest in perception led him to think about the process of discovery: we see what we are prepared to see, thus we must prepare ourselves to see more than we think possible. From this productive interplay of knowledge and a kind of artful ignorance, Thoreau derived his oft-repeated advice to "learn science & then forget it" (22 April 1852). Scientific knowledge is necessarily hypothetical: to hold on to a hypothesis too tightly is once again to squeeze out the unforeseen. As Thoreau commented early on, every "yea" must reserve a "nay for the morrow" (20 April 1840).

Thoreau's own "yea" to science reserved its "nay" for the moment when Spencer F. Baird invited him to join the American Association for the Advancement of Science. Thoreau declined, furiously in his *Journal*, courteously in his formal reply. To the one he complained that "they do not believe in a science which deals in higher law" (5 March 1853); to the other he gave the excuse that distance would prevent him from attending the meetings. Despite this episode, his actual relationship with science deepened: he bought a telescope and used it; he continued to read in science and to develop his own ideas of a science that *would* believe in "higher law." At Walden, Thoreau had surveyed the pond and sounded its depths. The lesson was narrated in *Walden*: the universe did indeed have a bottom that would hold an anchor. The knowledge that related the breadth and depth of Walden to the breadth and depth of a human soul was simultaneously physical and ethical, bespeaking a harmony that no single, individual view would ever compass entire. For his "nay," a deeper "yea."

After the publication of *Walden* in 1854, the *Journal* lapsed into a period of relative silence on matters scientific until 1856, when a neighbor's questions prompted Thoreau to look afresh at forests as plant communities that change over time. Whole new questions appeared: What are the patterns of change, and what causes them? How do human beings participate in these changes? He now saw the forest not as a passive repository of species but as an active network of "agents," starting with seeds, squirrels, and birds, whose myriad interactions wove a community in continual flux. Thoreau became a forest "geologist," counting and reading tree rings instead of strata, tracing patterns of change back into the past, and attending to the panorama offered by the fast-changing, fast-growing cutover lands around Concord. "My expectation ripens to discovery," he exulted (2 September 1856). Man is clearly no mere spectator in this panorama, but one participant among many, whose actions play unexpectedly into the desires of seedling oaks and nascent pines, and who is not nearly so important in the forest economy as "vermin" like jays and squirrels. Such creatures are not objects for research but our "living contemporar[ies]," and Thoreau's science would insist on acknowledging the necessary human connection, a connection more often seen in the antique herbals and natural histories than in "modern more scientific ones" (16 December 1859).

In "The Succession of Forest Trees" of September 1860, his one lecture on "a purely scientific subject," Thoreau demonstrated his ideal of science: rigorous and carefully researched, yet also passionately engaged with both the forests he loved and the audience who would determine their fate. In Thoreau's world, we can hardly afford to separate ourselves from nature and study it as if we bore no part in it; on the contrary, we are hip deep in our subject, busy altering the nature we pretend only to study and knowing ourselves only through the actions by which we pretend to know only the world. It was typical of Thoreau that he took his vision of an activist science not to the professionals at Cambridge and Boston but to the farmers and landowners of the Middlesex Agricultural Society, and spoke not in the dry and "objective" language of the scientific report but with warmth, humor, irony, and even a touch of despair. In a hybrid of science and sermon, Thoreau affirmed the power of his insights for rational forest management and lamented that the mass of men would rather watch the deceptive magic of the carnival than the real magic revealed by science: "Surely, men love darkness rather than light," he ended.

Yet his own continuing work belied this pessimism. "Succession" was to have been only the first piece of a grand new project to capture in language the "history" and "economy" of New England nature. The weeks that followed his public appearance record an explosion of activity, as Thoreau labored to "unroll the rotten papyrus on which the history of the Concord forest is written" (19 October 1860). Yet late that fall, just as the hard work of a decade was bearing fruit, Thoreau contracted a cold from which he never recovered; he died of tuberculosis in May 1862. His last journal entry records one final observation of nature's uncanny ability to write itself and his own canniness in learning to read the traces left by the interaction of wind, rain, and matter. Some idea of what his methods might have generated is given by the manuscripts he left behind at his death, "Wild Fruits" and "The Dispersion of Seeds."* Both manuscripts were books in embryo, and they developed through all the detail of nature the idea that most compelled Thoreau, the faith that drew him to science: "We find ourselves in a world that is already planted, but is also still being planted as at first" (18 October 1860). As one of his last *Journal* entries declares, "A seed, which is a plant or tree in embryo, which has the principle of growth, of life, in it, is more important in my eyes, and in the economy of Nature, than the diamond of Kohinoor" (22 March 1861). The material causes visible to us now, in the present, connect us across time to the deepest past and the most speculative future. All the world is connected and transported by the smallest facts of all: "Facts collected by a poet are set down at last as winged seeds of truth — samarae — tinged with his expectation. O may my words be verdurous & sempiternal as the hills. Facts fall from the poetic observer as ripe seeds" (19 June 1852).

And so Thoreau left us with the single fact that carries the greatest truth: the seed, his everlasting samara of thought.

---

* "The Dispersion of Seeds" has recently been published for the first time in *Faith in a Seed,* edited by Bradley P. Dean (Washington, D.C.: Island Press, 1993). Dean is currently preparing the longer manuscript of "Wild Fruits" for publication by Norton Press, forthcoming in 1999.

Material Faith

I

"The true man of science"

December 1837–November 1850

FACTS.

How indispensable to a correct study of nature is a perception of her true meaning— The fact will one day flower out into a truth. The season will mature and fructify what the understanding had cultivated. Mere accumulators of facts— collectors of materials for the masterworkmen, are like those plants growing in dark forests, which "put forth only leaves instead of blossoms."

16 December 1837, *Journal* 1:19

Men are constantly dinging in my ears their fair theories and plausible solutions of the universe— but ever there is no help— and I return again to my shoreless-islandless ocean, and fathom unceasingly for a bottom that will hold an anchor, that it may not drag.

13 August 1838, *Journal* 1:51

Linnaeus setting out for Lapland, surveys his "comb" and "spare shirt," "leather breeches," and "gauze cap to keep off gnats," with as much complacency as Buonaparte would a park of artillery to be used in the Russian

Campaign— His eye is to take in fish, flower, and bird, quadruped and biped— The quiet bravery of the man is admirable. These facts have even a *novel* interest.

22 November 1839, *Journal* 1:86

Science is always brave, for to know is to know good; doubt and danger quail before her eye. What the coward overlooks in his hurry, she calmly scrutinizes, breaking ground like a pioneer for the array of arts in her train.— Cowardice is unscientific—for there cannot be a science of ignorance— There may be a science of war—for that advances—but a retreat is rarely well conducted, if it is— then is it an orderly advance in the face of circumstances.

December 1839, *Journal* 1:91–92

A very meagre natural history suffices to make me a child —only their names and genealogy make me love fishes. I would know even the number of their fin rays—and how many scales compose the lateral line. I fancy I am amphibious and swim in all the brooks and pools in the neighborhood, with the perch and bream, or doze under the pads of our river amid the winding aisles and corridors formed by their stems, with the stately pickerel.

I am the wiser in respect to all knowledges, and the better qualified for all fortunes, for knowing that there is a minnow in the brook— Methink I have need even of his sympathy—and to be his fellow in a degree— I do

like him sometimes when he balances himself for an hour over the yellow floor of his basin.

I learned to-day that my ornithology had done me no service— The birds I heard, which fortunately did not come within the scope of my science — sung as freshly as if it had been the first morning of creation, and had for background to their song an untrodden wilderness — stretching through many a Carolina and Mexico of the soul.

The infinite bustle of nature of a summer's noon, or her infinite silence of a summer's night — gives utterance to no dogma. They do not say to us even with a seer's assurance, that this or that law is immutable — and so ever and only can the universe exist. But they are the indifferent occasion for all things — and the annulment of all laws.

The universe will not wait to be explained. Whoever seriously attempts a theory of it is already behind his age. His yea has reserved no nay for the morrow.

The era of greatest change is to the subject of it the condition of greatest invariableness. The longer the lever the less perceptible its motion. It is the slowest pulsation which is the most *vital*. I am independent of the change I detect.

My most essential progress must be to me a state of absolute rest. So in geology we are nearest to discovering the true causes of the revolutions of the globe, when we allow them to consist with a quiescent state of the elements. We discover the causes of all past change in the present invariable order of the universe.

The pulsations are so long that in the interval there is almost a stagnation of life. The first cause of the universe makes the least noise. Its pulse has beat but once — is now beating. The greatest appreciable revolutions are the work of the light-footed air — the stealthy-paced water — and the subterranean fire. The wind makes the desert without a rustle.

To every being consequently its own first cause is an insensible and inconceivable agent.

18 October 1840, *Journal* 1:190–91

Nov. 11th I obtained a levelling instrument and circumferentor combined, and have since ascertained the height of the cliff hill — and surveyed other objects.

After 15 November 1840, *Journal* 1:197

Nature is right, but man is straight. She erects no beams, she slants no rafters, and yet she builds stronger and truer than he. Every where she preaches not abstract but practical truth — She is no beauty at her toilet, but her cheek is flushed with exercise. The moss grows over her triangles. Unlike the man of science she teaches that skeletons are only good to wear the flesh, and make fast the sinews to — that better is the man than his bones.

15 December 1840, *Journal* 1:204

The great thoughts of a wise man seem to the vulgar who do not generalize to stand far apart like isolated mounts —but Science knows that the mountains which rise so solitary in our midst are parts of a great mountain chain —dividing the earth— And the eye that looks into the horizon—toward the blue sierra melting away in the distance may detect their flow of thought—

24 March 1842, *Journal* 1:389

"He has something demoniacal in him"

These volumes deal much in measurements and minute descriptions, not interesting to the general reader, with only here and there a colored sentence to allure him, like those plants growing in dark forests, which bear only leaves without blossoms. But the ground was comparatively unbroken, and we will not complain of the pioneer, if he raises no flowers with his first crop. Let us not underrate the value of a fact; it will one day flower in a truth. It is astonishing how few facts of importance are added in a century to the natural history of any animal. The natural history of man himself is still being gradually written. Men are knowing enough after their fashion. Every countryman and dairy-maid knows that the coats of the fourth stomach of the calf will curdle milk, and what particular mushroom is a safe and nutritious diet. You cannot go into any field or wood, but it will seem as if every stone had been turned, and the bark on every tree ripped up. But, after all, it is much easier to discover than to see when the cover is off. It has been well said that "the attitude of inspection is prone." Wisdom does not inspect, but behold. We must look a long time before we can see.

Slow are the beginnings of philosophy. He has something demoniacal in him, who can discern a law or couple two facts. We can imagine a time when "Water runs down hill" may have been taught in the schools. The true man of science will know nature better by his finer organization; he will smell, taste, see, hear, feel, better than other men. His will be a deeper and finer experience. We do not learn by inference and deduction and the application of mathematics to philosophy, but by direct intercourse and sympathy. It is with science as with ethics, — we cannot know truth by contrivance and method; the Baconian is as false as any other, and with all the helps of machinery and the arts, the most scientific will still be the healthiest and friendliest man, and possess a more perfect Indian wisdom.

"Natural History of Massachusetts,"
The Dial, 2 July 1842 (*Excursions*, 160–62)

I hate museums, there is nothing so weighs upon the spirits. They are catacombs of Nature. They are preserved death. One green bud of Spring one willow catkin, one faint trill from some migrating sparrow, might set the world on its legs again.

24 September 1843, *Journal* 1:465

Sometimes we would fain see events as merely material — wooden — rigid — dead — but again we are reminded that we actually inform them with little life by which they live. That they are the slaves and creatures of our conduct.

When dull and sensual I believe these are cornstalks good for cattle no more nor less — The laws of nature are science but in an enlightened moment they are morality

and modes of divine life. In a medium intellectual state
they are aesthetics.

Men have a strange taste for death who prefer to go to
museums to behold the cast off garments of life—rather
than handle the life itself. Where is the proper Herbar-
ium—the cabinet of shells—the museum of skeletons
but in the meadow—where the flower bloomed—or by
the sea-side where tide cast up the fish—or on the hills
where the beast laid down his life. Where the skeleton of
the traveller reposes in the grass there may it profitably
be studied. What right has mortal man to parade any
skeleton on its legs when once the gods have unloosed its
sinews—what right to imitate heaven with his wires—or
to stuff the body with sawdust—which nature has de-
creed shall return to dust again?

All the fishes that swim in the ocean can hardly atone
for the wrong done by stuffing and varnishing and encas-
ing under glass the relics of one inhabitant of the deep.—
Go to Italy and Egypt if you would behold these things
where bones are the natural product of the soil which
bears tombs and catacombs. Would you live in a dried
specimen of a world? a pickled world. Embalming is a
sin against heaven and earth—against heaven who has
recalled the soul—and set free the servile elements—
against earth who is robbed of her dust.

I have had my right perceiving senses so disturbed in
these haunts as for long to mistake a veritable living man,
in the attitude of repose musing like myself as the place
requires, for a stuffed specimen. So are men degraded in
consequence.

There is all the refinement of civilized life in the woods, under a sylvan garb. The wildest scenes even have an air of domesticity and homeliness to the citizen— And when the flicker's cackle is heard in the clearings he is reminded that civilization has imported nothing into them. Science is welcome to their deepest recesses for there too nature obeys the same old civil laws.—

1842–44, Journal 2:21–22

All great laws are really known to the simple necessities of men before they become the subject of science.

1842–44, Journal 2:24

Facts must be learned directly and personally—but principles may be deduced from information. The collector of facts possesses a perfect physical organization—the philosopher a perfect intellectual one. One can walk— the other sit—one acts, the other thinks. But the poet in some degree does both and uses and generalizes the results of both—he generalizes the widest deductions of philosophy.

1842–44, Journal 2:53

Books are written with intentions and as part of a system — Books which contain the elements of knowledge— and the science of things—that is men's ignorance of things. elements which lead to human solutions—methods to a method—but never to a ray of absolute or divine knowledge. The scholar is taught the method of arriving at that dilemma in which men of science and philosophers—professors, now stand—and studies elements and the best classification with a view to this end solely.

Many a book is written which does not necessarily

suggest or imply the phenomenon or object to explain which it professes to have been written. But we may begin anywhere with nature. Strictly speaking there is no such thing as elementary knowledge There is always a chasm between knowledge and ignorance which the steps of science can never pass.

1842–44, Journal 2:91

When a man is warmed by the several modes I have described—what next does he want—not surely more warmth of the same kind—as more and nicer food—larger and more splendid houses—& the like—but to adventure into life—a little—his vacation having commenced— As science which is poetry *professed* by the civilized state—measuring the unfathomed with its telescope—& microscope—but feebly & partially—we want something more comprehensive & assertive which may be called con-science perhaps—and signify a practical growth—

Fall–Winter 1845–46, Journal 2:144
[at Walden Pond]

All material things are in some sense man's kindred, and subject to the same laws with him.

Even a taper is his relative—and burns not eternally, as some say of lamps found burning in ancient sepulchres—but only a certain number of his hours.

These things belong to the same dynasty or system of things. He witnesses their wasting and decay as well as his own   What mans experience does not embrace is to him stationary and eternal   Whether he wakes or sleeps the lamp still burns on and burns out—completing its life within his own.

He sees such objects at a very near angle. They have a very large parallax to him—but not so those tapers the fixed stars which are not both lit and burnt out in the life of a man—yet they too are his distant relations.

Fall 1846, *Journal* 2:354 [at Walden Pond]

Astronomy is that department of physics which answers to Prophesy the Seer's or Poets calling   It is a mild a patient deliberate and contemplative science. To see more with the physical eye than man has yet seen to see farther, and off the planet—into the system. Shall a man stay on this globe without learning something—without adding to his knowledge—merely sustaining his body and with morbid anxiety saving his soul. This world is not a place for him who does not discover its laws.

Dull Despairing and brutish generations have left the race where they found it or in deeper obscurity and night —impatient and restless ones have wasted their lives in seeking after the philosopher's stone and the elixir of life — These are indeed within the reach of science—but only of a universal and wise science to which an enlightened generation may one day attain. The wise will bring to the task patience humility (serenity)—joy—resolute labor and undying faith.

After 2 December 1846, *Journal* 2:359 [at Walden Pond]

"I send you 15 pouts, 17 perch, 13 shiners . . . all from the pond by my house"

Concord, June 1, 1847.

Dear Sir,—

I send you 15 pouts, 17 perch, 13 shiners, 1 larger land tortoise, and 5 muddy tortoises, all from the pond by my house. Also 7 perch, 5 shiners, 8 breams, 4 dace? 2 muddy tortoises, 5 painted do., and 3 land do., all from the river. One black snake, alive, and one dormouse? caught last night in my cellar. The tortoises were all put in alive; the fishes were alive yesterday, *i.e.*, Monday, and some this morning. Observe the difference between those from the pond, which is pure water, and those from the river.

I will send the light-colored trout and the pickerel with the longer snout, which is our large one, when I meet with them. I have set a price upon the heads of snapping turtles, though it is late in the season to get them.

If I wrote red-finned eel, it was a slip of the pen; I meant red-finned minnow. This is their name here; though smaller specimens have but a slight reddish tinge at the base of the pectorals.

Will you, at your leisure, answer these queries?

Do you mean to say that the twelve banded minnows which I sent are undescribed, or only one? What are the scientific names of those minnows which have any? Are the four dace I send to-day identical with one of the former, and what are they called? Is there such a fish as the black sucker described, — distinct from the common?

Letter to James Elliot Cabot, 1 June 1847,
Correspondence, 182–83

"Any eyes will see new worlds at once"

The eye which can appreciate the naked and absolute beauty of a scientific truth is far more rare than that which is attracted by a moral one. Few detect the morality in the former, or the science in the latter. Aristotle defined art to be . . . *the principle of the work without the wood*; but most men prefer to have some of the wood along with the principle; they demand that the truth be clothed in flesh and blood and the warm colors of life. They prefer the partial statement because it fits and measures them and their commodities best. But science still exists every where as the sealer of weights and measures at least.

We have heard much about the poetry of mathematics, but very little of it has yet been sung. The ancients had a juster notion of their poetic value than we. The most distinct and beautiful statement of any truth must take at last the mathematical form. We might so simplify the rules of moral philosophy, as well as of arithmetic, that one formula would express them both. All the moral laws are readily translated into natural philosophy, for often we have only to restore the primitive meaning of the words by which they are expressed, or to attend to their literal instead of their metaphorical sense. They are already *supernatural* philosophy. The whole body of what is now called moral or ethical truth existed in the golden age as abstract science. Or, if we prefer, we may say that the laws of Nature are the purest morality. The Tree of Knowledge is a Tree of Knowledge of good and evil. He is not a true man of science who does not bring some sympathy to his studies, and expect to learn something by behavior as well as by application. It is childish to rest in the discovery of mere coincidences, or of partial and extraneous laws. The study of geometry is a petty and idle exercise of the mind, if it is applied to no larger system than the starry one. Mathematics should be mixed not only with physics but with ethics, *that* is *mixed* mathematics. The fact which interests us most is the life of the naturalist. The purest science is still biographical. Nothing will dignify and elevate science while it is sundered so wholly from the moral life of its devotee, and he professes another religion than it teaches, and worships at a foreign shrine. Anciently the faith of a philosopher was identical with his system, or, in other words, his view of the universe.

My friends mistake when they communicate facts to me with so much pains. Their presence, even their exaggerations and loose statements, are equally good facts for me. I have no respect for facts even except when I would use them, and for the most part I am independent of those which I hear, and can afford to be inaccurate, or, in other words, to substitute more present and pressing facts in their place.

The poet uses the results of science and philosophy, and generalizes their widest deductions.

The process of discovery is very simple. An unwearied and systematic application of known laws to nature, causes the unknown to reveal themselves. Almost any *mode* of observation will be successful at last, for what is most wanted is method. Only let something be determined and fixed around which observation may rally. How many new relations a foot-rule alone will reveal, and to how many things still this has not been applied! What wonderful discoveries have been, and may still be, made, with a plumb-line, a level, a surveyor's compass, a thermometer, or a barometer! Where there is an observatory and a telescope, we expect that any eyes will see new worlds at once. I should say that the most prominent scientific men of our country, and perhaps of this age, are either serving the arts and not pure science, or are performing faithful but quite subordinate labors in particular departments. They make no steady and systematic approaches to the central fact. A discovery is made, and at once the attention of all observers is distracted to that, and it draws many analogous discoveries in its train; as if their work were not already laid out for them, but they had been lying on their oars. There is wanting constant and accurate observation with enough of theory to direct and discipline it.

But above all, there is wanting genius. Our books of science, as they improve in accuracy, are in danger of losing the freshness and vigor and readiness to appreciate the real laws of Nature, which is a marked merit in the oft-times false theories of the ancients. I am attracted by the slight pride and satisfaction, the emphatic and even exaggerated style in which some of the older naturalists speak of the operations of Nature, though they are better qualified to appreciate than to discriminate the facts. Their assertions are not without value when disproved. If they are not facts, they are suggestions for Nature herself to act upon. "The Greeks," says Gesner, "had a common proverb (Λαγος καθευδον) a sleeping hare, for a dissembler or counterfeit, because the hare sees when she sleeps; for this is an admirable and rare work of Nature, that all the residue of her bodily parts take their rest, but the eye standeth continually sentinel."

Observation is so wide awake, and facts are being so rapidly added to the sum of human experience, that it appears as if the theorizer would always be in arrears, and were doomed forever to arrive at imperfect conclusions; but the power to perceive a law is equally rare in all ages of the world, and depends but little on the number of facts observed. The senses of the savage will furnish him with facts enough to set him up as a philosopher. The ancients can still speak to us with authority, even on the themes of geology and chemistry, though these studies are thought to have had their birth in modern times. Much is said about the progress of science in these centuries. I should say that the useful results of science had accumulated, but that there had been no accumulation of knowledge, strictly speaking, for posterity; for knowledge is to be acquired only by a corresponding experience. How can we *know* what we are *told* merely? Each man can interpret another's experience only by his own. We read that Newton discovered the law of gravitation, but how many who have heard of his famous discovery have recognized the same truth that he did? It may be not one. The revelation which was then made to him has not been superseded by the revelation made to any successor.—

> We see the *planet* fall,
> And that is all.

In a review of Sir James Clark Ross' Antarctic Voyage of Discovery, there is a passage which shows how far a body of men are commonly impressed by an object of sublimity, and which is also a good instance of the step from the sublime to the ridiculous. After describing the discovery of the Antarctic Continent, at first seen a hundred miles distant over fields of ice, — stupendous ranges of mountains from seven and eight to twelve and fourteen thousand feet high, covered with eternal snow and ice, in solitary and inaccessible grandeur, at one time the weather being beautifully clear, and the sun shining on the icy landscape; a continent whose islands only are accessible, and these exhibited "not the smallest trace of vegetation," only in a few places the rocks protruding through their icy covering, to convince the beholder that land formed the nucleus, and that it was not an iceberg; —the practical British reviewer proceeds thus, sticking to his last, "On the 22d of January, afternoon, the Expedition made the latitude of 74° 20', and by 7^h P. M., having ground (ground! where did they get ground?) to believe that they were then in a higher southern latitude than had been attained by that enterprising seaman, the late Captain James Weddel, and therefore higher than all their predecessors, an extra allowance of grog was issued to the crews as a reward for their perseverance."

Let not us sailors of late centuries take upon ourselves any airs on account of our Newtons and our Cuviers. We deserve an extra allowance of grog only.

"Friday," *A Week on the Concord and*
Merrimack Rivers, 361–66

The scientific startling & successful as it is, is always some thing less than the vague poetic—it is that of it which subsides—it is the sun shorn of its beams a mere disk—the sun indeed—but—no longer phosphor— light bringer or giver—or εχαεργοξ far working. Science applies a finite rule to the infinite. — & is what you can weigh & measure and bring away. Its sun no longer dazzles us and fills the universe with light.

5 January 1850, *Journal* 3:44

Nothing memorable was ever accomplished in a prosaic mood of mind. The heroes & discoverers have found true more than was previously believed only when they were expecting & dreaming of something more than their contemporaries dreamed of—when they were in a frame of mind prepared in some measure for the truth

Referred to the world's standard—the hero, the discoverer—is insane. its greatest men are all insane. At first the world does not respect its great men— Some rude and simple nations go to the other extreme & reverence all kinds of insanity. — Humboldt says—speaking of Columbus, approaching the New World—"The grateful coolness of the evening air, the etherial purity of the starry firmament, the balmy fragrance of flowers, wafted to him by the land breeze—all led him to suppose . . . that he was approaching the garden of Eden, the sacred abode of our first parents. The Orinoco seemed to him one of the four rivers, which according to the venerable tradition of the ancient world, flowed from Paradise, to water & divide the surface of the earth, newly adorned with plants."

12 May 1850, *Journal* 3:67

II

"How many questions there are!"

November 1850–April 1852

Apparently all but the evergreens & oaks have lost their leaves now. It is singular that the shrub-oaks retain their leaves through the winter, why do they?

16 November 1850, *Journal* 3:142

The first really cold day. I find on breaking off a shrub-oak leaf a little life at the foot of the leaf-stalk so that a part of the green comes off— It has not died quite down to the point of separation as it will do, I suppose, before spring. Most of the oaks have lost their leaves except on the lower branches, as if they were less exposed and less mature there and felt the changes of the seasons less.

19 November 1850, *Journal* 3:145

Horace Hosmer was picking out today half a bushel or more of a different & better kind of cranbery as he thought, separating them from the rest— . . . I must see him about it. It may prove to be one more of those instances in which the farmer detects a new species—and makes use of the knowledge from year to year in his profession while the botanist expressly devoted to such investigations has failed to observe it. . . .

It is often the unscientific man who discovers the new species — It would be strange if it were not so. But we are accustomed properly to call that only a scientific discovery which knows the relative value of the thing discovered — uncovers a fact to mankind.

20 November 1850, Journal 3:146–47

The knowledge of an unlearned man is living & luxuriant like a forest — but covered with mosses & lichens and for the most part inaccessible & going to waste — the knowledge of the man of science is like timber collected in yards for public works which stub supports a green sprout here & there — but even this is liable to dry rot.

7 January 1851, Journal 3:174

You might say of a very old & withered man or woman that they hang on like a shrub-oak leaf almost to a second spring. There was still a little life in the heel of the leaf-stalk

8 January 1851, Journal 3:175

I have heard that there is a Society for the Diffusion of Useful Knowledge — It is said that Knowledge is power and the like —

Methinks there is equal need of a society for the diffusion of useful Ignorance — for what is most of our boasted so called knowledge but a conceit that we know something which robs us of the advantages of our actual ignorance — ...

For a man's ignorance sometimes is not only useful but beautiful while his knowledge is oftentimes worse than useless beside being ugly. ...

My desire for knowledge is intermittent but my desire

to commune with the spirit of the universe—to be intoxicated even with the fumes, call it, of that divine nectar—
to bear my head through atmospheres and over heights
unknown to my feet—is perennial & constant.

9 February 1851, *Journal* 3:184-85

Obey the law which reveals and not the law revealed.

27 February 1851, *Journal* 3:201

It is a certain faery land where we live—you may walk
out in any direction over the earth's surface—lifting your
horizon—and every where your path—climbing the
convexity of the globe leads you between heaven & earth
— —not away from the light of the sun & stars—& the
habitations of men. I wonder that I even get 5 miles on
my way—the walk is so crowded with events—& phenomena. How many questions there are which I have not
put to the inhabitants!

7 June 1851, *Journal* 3:245

I have been tonight with Anthony Wright to look through
Perez Bloods Telescope a 2nd time. A dozen of *his*
Bloods neighbors were swept along in the stream of our
curiosity. . . . I was amused to see what sort of respect this
man with a telescope had obtained from his neighbors—
something akin to that which savages award to civilized
men—though in this case the interval between the parties was very slight. Mr Blood with his scull cap on his
short figure—his north European figure made me think
of Tycho Brahe— He did not invite us into his house
this cool evening—men nor women— Nor did he ever
before to my knowledge

I am still contented to see the stars with my naked eye
Mr Wright asked him what his instrument cost   He answered — "Well, that is something I dont like to tell. (stut-
tering or hesitating in his speech a little, as usual) It is a
very proper question however" — "Yes," said I, "and you
think that you have given a very proper answer."

7 July 1851, *Journal* 3:289

Visited the Observatory. Bond said they were catalogu-
ing the stars at Washington? or trying to. They do not at
Cambridge[—]of no use with their force. Have not force
enough now to make mag. obs. When I asked if an observer with the small telescope could find employment —
he said "O yes — there was employment enough for observation with the naked eye — observing the changes in
the brilliancy of stars &c &c — if they could only get
some good observers. —["] One is glad to hear that the
naked eye still retains some importance in the estimation
of astronomers.

9 July 1851, *Journal* 3:296–97

For years I marched as to a music in comparison with
which the military music of the streets is noise & dis-
cord. I was daily intoxicated and yet no man could call
me intemperate. With all your science can you tell how it
is — & whence it is, that light comes into the soul?

16 July 1851, *Journal* 3:306

The season of morning fogs has arrived I think it is connected with dog days   Perhaps it is owing to the greater
contrast between the night & the day — the nights being
nearly as cold while the days are warmer? Before I rise

from my couch I see the ambrosial fog stretched over the river draping the trees — ... it is almost musical; it is positively fragrant. How faery like it has visited our fields. I am struck by its firm outlines as distinct as a pillow's edge about the height of my house — a great crescent over the course of the river from SW to NE.

5½ Am   Already some parts of the river are bare — It goes off in a body down the river before this air — and does not rise into the heavens — It retreats & I do not see how it is dissipated. This slight thin vapor which is left to curl over the surface of the still dark water still as glass — seems not [to] be the same things — of a different quality.

I hear the cockrils crow through it — and the rich crow of young roosters — that sound indicative of the bravest rudest health — hoarse without cold — hoarse with a rude health   That crow is all nature compelling — famine & pestilence flee before it — These are our fairest days which are born in a fog

I saw the tall lettuce yesterday Lactucca elongata — whose top or main shoot had been broken off — & it had put up various stems — with entire & lanceolate — not runcinate leaves as usual — thus making what some botanists have called a variety — β. linearis — So I have met with some Geniuses who having met with some such accident maiming them — have been developed in some such *monstrous* & partial though original way. They were original in being less than themselves.

22 July 1851, *Journal* 3:326-27

The mind is subject to moods as the shadows of clouds pass over the earth — Pay not too much heed to them — Let not the traveller stop for them — They consist with the fairest weather. By the mood of my mind I suddenly felt dissuaded from continuing my walk—but I observed at the same instant that the shadow of a cloud was passing one spot on which I stood—though it was of small extent—which if it had no connexion with my mood at any rate suggested how transient & little to be regarded that mood was— I kept on & In a moment the sun shone on my walk within & without. . . .

But this habit of close observation— In Humboldt—Darwin & others. Is it to be kept up long—this science — Do not tread on the heels of your experience Be impressed without making a minute of it. Poetry puts an interval between the impression & the expression—waits till the seed germinates naturally.

23 July 1851, *Journal* 3:331

I went to a menagerie the other day. The proprietors had taken wonderful pains to collect rare and interesting animals from all parts of the world. And then placed by them —a few stupid and ignorant fellows who knew little or nothing about the animals & were unwilling even to communicate the little they knew. . . . Why all this pains taken to catch in Africa—and no pains taken to exhibit in America? Not a cage was labelled— There was nobody to tell us how or where the animals were caught—or what they were— Probably the proprietors themselves do not know—or what their habits are— But hardly had we been ushered into the presence of this choice this admirable collection—than a ring was formed for Master Jack & the poney. . . .

Why I expected to see some descendant of Cuviers there to improve this opportunity for a lecture on Nat. Hist.

That is what they should do make this an—occasion

for communicating some solid information— ...I go not there to see a man hug a lion—or fondle a tiger—but to learn how he is related to the wild beast— ...At present foolishly the professor goes alone with his poor painted illustrations of animated— While the menagerie takes another road without its professor only its keepers.

After 1 August 1851, *Journal* 3:350–51

Ah what a poor dry compilation is the Annual of Scientific Discovery. I trust that observations are made during the year which are not chronicled there. That some mortal may have caught a glimpse of Nature in some corner of the earth during the year—1851. One sentence of Perennial poetry would make me forget—would atone for volumes of mere science. The astronomer is as blind to the significant phenomena—or the significance of phenomena as the wood-sawyer who wears glasses to defend his eyes from sawdust— The question is not what you look at—but how you look & whether you see.

5 August 1851, *Journal* 3:354–55

The poet must be continually watching the moods of his mind as the astronomer watches the aspects of the heavens. What might we not expect from a long life faithfully spent in this wise—the humblest observer would see some stars shoot. ... A meteorological journal of the mind— You shall observe what occurs in your latitude, I in mine. ...

The grass in the high pastures is almost as dry as hay — The seasons do not cease a moment to revolve and therefore nature rests no longer at her culminating point than at any other. If you are not out at the right instant the

summer may go by & you not see it. How much of the year is spring & fall—how little can be called summer! The grass is no sooner grown than it begins to wither—
. . . .

The most inattentive walker can see how the science of geology took its rise. The inland hills & promontories betray the action of water on their rounded sides as plainly as if the work were completed yesterday. He sees it with but half an eye as he walks & forgets his thought again. Also the level plains & more recent meadows & marine shells found on the tops of hills— The Geologist painfully & elaborately follows out these suggestions— & hence his fine spun theories. . . .

I fear that the character of my knowledge is from year to year becoming more distinct & scientific— That in exchange for views as wide as heaven's cope I am being narrowed down to the field of the microscope— I see details not wholes nor the shadow of the whole. I count some parts, & say 'I know'. The cricket's chirp now fills the air in dry fields near pine woods.

19 August 1851, *Journal* 3:377–80

I hear a cricket in the depot field—walk a rod or two and find the note proceeds from near a rock— Partly under a rock between it & the roots of the grass he lies concealed—for I pull away the withered grass with my hands—uttering his night-like creak with a vibratory motion of his wings & flattering himself that it is night because he has shut out the day— He was a black fellow nearly an inch long with two long slender feelers.... Every milkman has heard them all his life—it is the sound that fills his ears as he drives along—but what one has ever got off his cart to go in search of one? I see smaller ones moving stealthily about whose note I do not know   Who ever distinguished their various notes? which fill the crevices in each others song— It would be a curious ear indeed that distinguished the species of the crickets which it heard—& traced even the earth song home each part to its particular performer   I am afraid to be so knowing....

How copious & precise the botanical language to describe the leaves, as well as the other parts of a plant. Botany is worth studying if only for the precision of its terms—to learn the value of words & of system. It is wonderful how much pains has been taken to describe a flowers leaf—, compared for instance with the care that is taken in describing a psychological fact. Suppose as much ingenuity (perhaps it would be needless) in making a language to express the sentiments. We are armed with language adequate to describe each leaf in the field. — or at least to distinguish it from each other—but not to describe a human character—with equally wonderful indistinctness & confusion we describe men— The precision and copiousness of botanical language applied to the description of moral qualities! ...

A traveller who looks at things with an impartial eye may see what the oldest inhabitant has not observed.

20 August 1851, *Journal* 3:381–84

What a faculty must that be which can paint the most barren landscape and humblest life in glorious colors It is pure & invigorated senses reacting on a sound & strong imagination. Is not that the poets case? The intellect of most men is barren. They neither fertilize nor are fertilized. It is the marriage of the soul with nature that makes the intellect fruitful — that gives birth to imagination. When we were dead & dry as the high-way some sense which has been healthily fed will put us in relation with nature in sympathy with her — some grains of fertilizing pollen floating in the air fall on us — & suddenly the sky is all one rain bow — is full of music & fragrance & flavor — The man of intellect only the prosaic man is a barren & staminiferous flower the poet is a fertile & perfect flower Men are such confirmed arithmeticians & slaves of business that I cannot easily find a blank book that has not a red line or a blue one for the dollars and cents, or some such purpose. . . .

Any anomaly in vegetation makes nature seem more real & present in her working — as the various red & yellow excrescences on young oaks — I am affected as if it were a different Nature that produced them. As if a poet were born — who had designs in his head. . . .

Hieracium Paniculatum a very delicate & slender hawkweed — I have now found all the hawkweeds. Singular these genera of plants — plants manifestly related yet distinct — They suggest a history to Nature — a Natural *history* in a new sense.

21 August 1851, *Journal* 4:3–6

We cannot write well or truly but what we write with gusto. The body the senses must conspire with the spirit — Expression is the act of the whole man. that our speech may be vascular— The intellect is powerless to express thought without the aid of the heart & liver and of every member— Often I feel that my head stands out too dry—when it should be immersed. A writer a man writing is the scribe of all nature—he is the corn & the grass & the atmosphere writing. It is always essential that we love to do what we are doing—do it with a heart. The maturity of the mind however may perchance consist with a certain dryness.

2 September 1851, *Journal* 4:27–28

The man of science discovers no world for the mind of man with all its faculties to inhabit—Wilkinson finds a *home* for the imagination—& it is no longer out cast & homeless. All perception of truth is the detection of an analogy.— we reason from our hands to our head.

5 September 1851, *Journal* 4:46

I seem to be more constantly merged in nature—my intellectual life is more obedient to nature than formerly— but perchance less obedient to Spirit— I have less memorable seasons. I exact less of myself. I am getting used to my meanness—getting to accept my low estate— O if I could be discontented with myself! If I could feel anguish at each descent!

12 October 1851, *Journal* 4:141

It is a rare qualification to be ably to state a fact simply & adequately. To digest some experience cleanly. To say yes and no with authority — To make a square edge. To conceive & suffer the truth to pass through us living & intact — even as a waterfowl an eel — thus peopling new waters. First of all a man must see, before he can say. — Statements are made but partially — Things are said with reference to certain conventions or existing institutions. — not absolutely. A fact truly & absolutely stated is taken out of the region of commonsense and acquires a mythologic or universal significance. Say it & have done with it. Express it without expressing yourself. See not with the eye of science — which is barren — nor of youthful poetry which is impotent. But taste the world. & digest it. It would seem as if things got said but rarely & by chance — As you *see* so at length will you *say*. . . .

At first blush a man is not capable of reporting truth — he must be drenched & saturated with it first. What was *enthusiasm* in the young man must become *temperament* in the mature man. without excitement — heat or passion he will survey the world which excited the youth — & threw him off his balance. As all things are significant; so all words should be significant. . . .

This on my way to Conantum 2½ Pm It is a bright clear warm november day. I feel blessed. I love my life. I warm toward all nature.

1 November 1851, Journal 4:157–59

In our walks C. takes out his note-book some times & tries to write as I do — but all in vain. He soon puts it up again — or contents himself with scrawling some sketch of the landscape. Observing me still scribbling he will say that *he* confines himself to the ideal — purely ideal remarks—he leaves the facts to me. Sometimes too he will say a little petulantly—"*I* am universal I have nothing to do with the particular and definite." . . .

I too would fain set down something beside facts. Facts should only be as the frame to my pictures— They should be material to the mythology which I am writing. Not facts to assist men to make money—farmers to farm profitably in any common sense. Fact to tell who I am—and where I have been—or what I have thought. . . . My facts shall all be falsehoods to the common sense. I would so state facts that they shall be significant shall be myths or mythologic. Facts which the mind perceived—thoughts which the body thought with these I deal—

9 November 1851, *Journal* 4:170

The chopper who works in the woods all day for many weeks or months at a time becomes intimately acquainted with them in his way. He is more open in some respects to the impressions they are fitted to make than the naturalist who goes to see them He is not liable to exaggerate insignificant features. He really forgets himself—forgets to observe—and at night he *dreams* of the swamp its phenomena & events. Not so the naturalist; enough of his unconscious life does not pass there. . . .

You must be conversant with things for a long time to know much about them—like the moss which has hung from the spruce—and as the partridge & the rabbit are acquainted with the thickets & at length have acquired the color of the places they frequent. If the man of science can put all his knowledge into propositions—the wood man has a great deal of incommunicable knowledge.

18 November 1851, *Journal* 4:192–93

I witness a beauty in the form or coloring of the clouds which addresses itself to my imagination — for which you account scientifically to my understanding — but do not so account to my imagination. It is what it suggests & is the symbol of that I care for — and if by any trick of science you rob it of its symbolicalness you do me no service & explain nothing. . . .

What sort of science is that which enriches the understanding but robs the imagination. Not merely robs Peter to pay Paul — but takes from Peter more than it ever gives to Paul.

That is simply the way in which it speaks to the understanding and that is the account which the understanding gives of it — but that is not the way it speaks to the Imagination & that is not the account which the Imagination gives of it. Just as inadequate to a pure mechanic would be a poets account of a steam engine.

If we knew all things thus mechanically merely — should we know anything really? . . .

Ah give me pure mind — pure thought. Let me not be in haste to detect the *universal law*, let me see more clearly a particular instance. Much finer themes I aspire to — which will yield no satisfaction to the vulgar mind — not one sentence for them — Perchance it may convince such that there are more things in heaven & earth than are dreamed of in their philosophy. Dissolve one nebula — & so destroy the nebular system & hypothesis. Do not seek expressions — seek thoughts to be expressed. By perseverance you get two views of the same rare truth.

25 December 1851, *Journal* 4:221–23

Men talk about travelling this way or that as if seeing were all in the eyes, and a man could sufficiently report what he stood bodily before. — When the seeing depends ever on the living. All report of travel is the report of victory or defeat—of a contest with every event and phenomenon & how you came out of it. A blind man who possesses inward truth & consistency will see more than one who has faultless eyes but no serious & laborious astronomer to look through them.

12 January 1852, *Journal* 4:249–50

But after all where is the flower lore—for the first book & not the last should contain the poetry of flowers— The natural system may tell us the value of a plant in medicine or the arts or for food—but neither it nor the Linnaean to any great extent tell us its chief value & significance to man—which in any measure accounts for its beauty.— its flower like properties— There will be pages about some fair flower's qualities as food or medicine—but perhaps not a sentence about its significance to the eye—as if the cowslip were better for greens than for yellows. Not about what children & all flower lovers gather flowers for—are they emisaries sent forth by the arts to purvey & explore for them? . . . What a keep-sake a manual of botany! In which is uttered breathed man's love of flowers. It is dry as a hortus siccus.— Flowers are pressed into the botanist's service. . . .

There must be the copulating & generating force of love behind every effort destined to be successful. The cold resolve gives birth to—begets nothing. The theme that seeks me, not I it. The poet's relation to this theme is the relation of lovers. It is no more to be courted. Obey —report.

30 January 1852, *Journal* 4:306–7

Botanies instead of being the poetry are the prose of flowers. I do not mean to underrate Linnaeu's admirable nomenclature much of which is itself poetry.

31 January 1852, *Journal* 4:310

The audience are never tired of hearing how far the wind carried some man woman or child — or family bible — but they are immediately tired if you undertake to give them a scientific account of it.

4 February 1852, *Journal* 4:328

I suspect that the child plucks its first flower with an insight into its beauty & significance which the subsequent botanist never retains.

5 February 1852, *Journal* 4:329

The artificial system has been very properly called the dictionary — and the natural method, the, grammar of the science of botany, by botanists themselves — But are we to have nothing but grammars and dictionaries in this literature? Are there no works written in the language of the flowers?

I asked a learned & accurate naturalist who is at the same time the courteous guardian of a public library to direct me to those works which contained the more particular — *popular* account or *biography* of particular flowers from which the botanies I had met with appeared to draw sparingly — for I trusted that each flower had had many lovers & faithful describers in past times — but he informed me that I had read all — that no one was acquainted with them — they were only catalogued like his books.

6 February 1852, *Journal* 4:330

The actual bee-hunter—and pigeon-catcher—is familiar with facts in the natural history of bees & pigeons—
which Huber & even Audubon are totally ignorant of. I
love best the unscientific man's knowledge there is so
much more humanity in it. It is connected with true
sports. . . .

Color, which is the poets wealth is so expensive, that
most take to mere outline or pencil sketches & become
men of science.

13 February 1852, *Journal* 4:345–47

By the artificial system we learn the names of plants—by
the natural their relations to one another—but still it remains to learn their relation to man— The poet does
more for us in this department.

16 February 1852, *Journal* 4:353

If you would read books on botany go to the fathers of
the science— Read Linnaeus at once, & come down
from him as far as you please— I lost much time reading
the Florists. It is remarkable how little the mass of those
interested in botany are acquainted with Linnaeus. His
Philosophia Botanica which Rousseau Sprengel & others praised so highly—I doubt if it has ever been translated into English.— It is simpler more easy to under-

stand & more comprehensive — than any of the hundred manuals to which it has given birth — A few pages of cuts representing the different parts of plants with the botanical names attached — is worth whole volumes of explanation.

According to Linnaeu's classification, I come under the head of the *Miscellaneous* Botanophilists.

17 February 1852, *Journal* 4:354

I have a common place book for facts and another for poetry — but I find it difficult always to preserve the vague distinction which I had in my mind — for the most interesting & beautiful facts are so much the more poetry and that is their success. They are *translated* from earth to heaven — I see that if my facts were sufficiently vital & significant — perhaps transmuted more into the substance of the human mind — I should need but one book of poetry to contain them all.

PM TO FAIR HAVEN HILL

One discovery in Meteorology, one significant observation is a good deal. I am grateful to the man who introduces order among the clouds. Yet I look up into the heavens so fancy free, I am almost glad not to know any law for the winds....

It is impossible for the same person to see things from the poet's point of view and that of the man of science. The poets second love may be science — not his first. — when use has worn off the bloom. I realize that men may be born to a condition of mind at which others arrive in middle age by the decay of their poetic faculties.

18 February 1852, *Journal* 4:356–57

However I can see that there is a certain advantage in these hard & precise terms—such as the lichenist uses for instance— No one masters them so as to use them in writing on the subject without being far better informed than the rabble about it. New books are not written on chemistry—or cryptogamia of as little worth comparitively as are written on the *spiritual* phenomena of the day.

No man writes on lichens using the terms of the science intelligibly—without having something to say—but every one thinks himself competent to write on the relation of the soul to the body as if that were a *phaenogamous* subject.

1 March 1852, Journal 4:368–69

If the sciences are protected from being carried by assault by the mob by a palisade or chevaux de frise of technical terms—so also the learned man may sometimes ensconce himself & conceal his little true knowledge behind hard names— Perhaps the value of any statement may be measured by its susceptibility to be expressed in popular language. The greatest discoveries can be reported in the newspapers.— I thought it was a great advantage both to speakers & hearers when at the meetings of scientific gentlemen at the Marlboro chapel—the representatives of all departments of science were required to speak intelligibly to those of other departments—therefore dispensing with the most peculiarly technical terms— A man may be permitted to state a very meager truth to a fellow student using technical terms—but when he stands up before the mass of men he must have some distinct & important truth to communicate—and

the most important it will always be the most easy to
communicate to the vulgar.

2 March 1852, Journal 4:369–70

I find myself inspecting little granules as it were on the
bark of trees—little shields or apothecia springing from
a thallus—such is the mood of my mind—and I call it
studying lichens. That is merely the prospect which is af-
forded me. It is short commons & innutritious. Surely I
might take wider views. The habit of looking at things
microscopically as the lichens on the trees & rocks really
prevents my seeing aught else in a walk. Would it not be
noble to study the shield of the sun—on the thallus of
the sky? cerulean, which scatters its infinite sporules of
light through the universe. To the lichenist is not the
shield (or rather the apothecia) of a lichen dispropor-
tionately large compared with the universe? The minute
appothecia of the Pertusaria which the woodchopper
never detected occupies so large a space in my eye at present as to shut out a great part of the world.

5 March 1852, Journal 4:377–78

I am glad to hear that naked eyes are of any use, for I cannot afford to buy a Munich (?) Telescope

20 March 1852, Journal 4:395

Too late now for the morning influence & inspiration.—
The birds sing not so earnestly & joyously—there is a
blurring ripple on the surface of the lake.— How few
valuable observations can we make in youth— What if
there were united the susceptibility of youth with the discrimination of age. Once I was part and parcel of nature
—now I am observant of her....

It appears to me that to one standing on the heights of philosophy mankind & the works of man will have sunk out of sight altogether. Man is altogether too much insisted on. The poet says the proper study of mankind is man— I say study to forget all that—take wider views of the universe— That is the egotism of the race. . . .

Man is but the place where I stand & the prospect (thence) hence is infinite. it is not a chamber of mirrors which reflect me—when I reflect myself—I find that there is other than me. . . .

How novel and original must be each new mans view of the universe—for though the world is so old—& so many books have been written—each object appears wholly undescribed to our experience—each field of thought wholly unexplored— The whole world is an America—a *New World*. The fathers lived in a dark age —& throw no light on any of our subjects. The sun climbs to the zenith daily high over all literature & science—astronomy even concerns us worldlings only— but the sun of poetry & of each new child born into the planet has never been astronomized, nor brought nearer by a telescope. So it will be to the end of time. The end of the world is not yet. Science is young by the ruins of Luxor—unearthing the sphinx—or Ninevah—or between the pyramids.

2 April 1852, *Journal* 4:416–21

What a novel life to be introduced to a dead sucker floating on the water in the spring!— Where was it spawned pray? The sucker is so recent—so unexpected—so unrememberable so unanticipatable a creation— While so many institutions are gone by the board and we are despairing of men & of ourselves there seems to be life even in a dead sucker—whose fellows at least are alive. The world never looks more recent or promising—religion philosophy poetry—than when viewed from this point. To see a sucker tossing on the spring flood—its swelling imbricated breast heaving up a bait to not despairing gulls— It is a strong & a strengthening sight. Is the world coming to an end?— Ask the chubs. As long as fishes spawn—glory & honor to the cold blooded who despair—as long as ideas are expressed—as long as friction makes bright—as long as vibrating wires make music of harps—we do not want redeemers.

15 April 1852, *Journal* 4:450

The sight of the sucker floating on the meadow at this season affects me singularly. as if it were a fabulous or mythological fish—realizing my *idea* of a fish— It reminds me of pictures of dolphins or of proteus. I see it for what it is—not an actual terrene fish—but the fair symbol of a divine idea—the design of an artist—its color & form—its gills & fins & scales—are perfectly beautiful —because they completely express to my mind what they were intended to express— It is as little fishy as a fossil fish. Such a form as is sculptured on ancient monuments and will be to the end of time.— made to point a moral. I am serene & satisfied when the birds fly & the fishes swim as in fable, for the moral is not far off. When the migration of the goose is significant and has a moral to it. When the events of the day have a mythological character & the most trivial is symbolical.

For the first time I perceive this spring that the year is a circle— I see distinctly the spring arc thus far. It is drawn with a firm line. Every incident is a parable of the great teacher. The cranberries washed up in the meadows & into the road on the causeways now yields a pleasant acid.

Why should just these sights & sounds accompany our life? Why should I hear the chattering of blackbirds —why smell the skunk each year? I would fain explore the mysterious relation between myself & these things. I would at least know what these things unavoidably are— —make a chart of our life—know how its shores trend— that butterflies reappear & when—know why just this circle of creatures completes the world. Can I not by expectation affect the revolutions of nature—make a day to bring forth something new?

As Cawley loved a garden, so I a forest. Observe all kinds of coincidences—as what kinds of birds come with what flowers.

18 April 1852, *Journal* 4:467–68

How sweet is the perception of a new natural fact!—suggesting that worlds remain to be unveiled. That phenomenon of the Andromeda seen against the sun cheers me exceedingly When the phenomenon was not observed— It was not—at all. I think that no man ever takes an original or detects a principle without experiencing an inexpressible as quite infinite & sane pleasure which advertises him of the dignity of that truth he has perceived.

The thing that pleases me most within these three days is the discovery of the andromeda phenomenon— It makes all those parts of the country where it grows more attractive & elysian to me. It is a natural magic. These little leaves are the stained windows in the cathe-

dral of my world. At sight of any redness I am excited like
a cow. —

19 April 1852, Journal 4:471

I want things to be incredible — too good to appear true.
C. says "after you have been to the P.O. once you are
dammed." — but I answer that it depends somewhat on
whether you get a letter or not. If you would be wise learn
science & then forget it.

22 April 1852, Journal 4:483

III

"I will not fear to know"

May 1852–August 1854

My dream frog turns out to be a toad. I watched half a dozen a long time at 3½ this afternoon in Hubbards pool, where they were frogging? lustily. They sat in the shade, either partly in the water, or on a stick, looked darker and narrower in proportion to their length than toads usually do, and moreover are aquatic— I see them jump into the ditches as I walk— ... One which I brought home answers well enough to the description of the Common toad bufo Americanus—though it is hardly so gray. Their piping (?) was evidently connected with their loves. Close by it is an unmusical monotonous deafening sound, a steady blast—(not a peep nor a croak—but a *kind* of piping)—but far away it is a dreamy lulling sound & fills well the crevices of nature.

6 May 1852, Journal 5:32-33

I fear that the dream of the toads will not sound so musical now that I know whence it proceeds. But I will not fear to *know*. They will awaken new & more glorious music for me as I advance—still farther in the horizon—not to be traced to toads & frogs in slimy pools.

7 May 1852, Journal 5:38

Up to about the 14th of May I watched the progress of the season very closely — though not so carefully the earliest birds — but since that date both from poor health & multiplicity of objects I have note[d] little but what fell under my observation.

19 May 1852, *Journal* 5:67

Facts collected by a poet are set down at last as winged seeds of truth — samarae — tinged with his expectation. O may my words be verdurous & sempiternal as the hills. Facts fall from the poetic observer as ripe seeds.

19 June 1852, *Journal* 5:112

I am inclined to think that my hat whose lining is gathered in mid way so as to make a shelf is about as good a botany box as I could have & far more convenient — and there is something in the darkness & the vapors that arise from the head — at least if you take a bath which preserves flowers through a long walk. Flowers will frequently come fresh out of this botany box at the end of the day though they have had no sprinkling. . . .

These are interesting groves of young soft white pines 18 feet high — whose vigorous yellowish green shoots of this season from 3 to 18 inches long at the extremities of all the branches contrast remarkably with the dark green of the old leaves. I observe that these shoots are bent and what is more remarkable all one way i.e. to the east — almost at a right angle the topmost ones — and I am reminded of the observation in Henry's Adventures that the Indians guided themselves in cloudy weather by this mark — All these shoots excepting those low down on the East side are bent toward the east. I am very much pleased with this observation confirming that of the Indians. . . . This gives me more satisfaction than any observation which I have made for a long time. This is true of the rapidly growing shoots. How long will this phenomenon avail to guide the traveller? How soon do they become erect? A natural compass. How few civilised men probably have ever made this observation—so important to the savage! How much may there have been known to his woodcraft—which has not been detected by science!

23 June 1852, Journal 5:126–29

I am disappointed to notice today that most of the pine tops incline to the west—as if the wind had to do with it.

24 June 1852, Journal 5:143

What a mean & wretched creature is man by & by some Dr Morton may be filling your cranium with white mustard seed to learn its internal capacity.

Of all ways invented to come at a knowledge of a living man—this seems to me the worst—as it is the most belated. You would learn more by once paring the toe nails of the living subject. There is nothing out of which the spirit has more completely departed—& in which it has left fewer significant traces.

25 June 1852, Journal 5:148–49

Science affirms too much. Science assumes to show *why* the lightning strikes a tree—but it does not show us the moral *why*, any better than our instincts did. It is full of presumption. Why should trees be struck? It is not enough to say because they are in the way. Science answers *non scio*—I am ignorant. All the phenomena of na-

ture need be seen from the point of view of wonder &
awe — like lightning — & on the other hand the lightning
itself needs to regarded with serenety as the most familiar
& innocent phenomena. There runs through the righteous man's moral spine — a rod with burnished points to
heaven which conducts safely away in to the earth the
flashing wrath of Nemesis — so that it merely clarifies the
air. This moment the confidence of the righteous man
erects a sure conductor within him — the next perchance
— a timid steeple diverts the fluid to his vitals.

If a mortal be struck with a thunder bolt *coelo sereno,*
it is naturally felt to be more aweful & vengeful. Men are
probably nearer to the essential truth in their supersti-
tions than in their science. Some places are thought to be
particularly exposed to lightning — some oaks on hill
tops for instance.

27 June 1852, Journal 5:159–60

Nature is reported not by him who goes forth con-
sciously as an observer — but in the fullness of life — to
such a one she rushes to make her report — To the full
heart she is all but a figure of speech. This is my year of
observation, & I fancy that my friends are also more devoted to outward observation than ever before — as if it
were an epidemic. I cross the brook by Hubbards little
bridge Now nothing but the cool invigorating scent
which is perceived at night in these low meadowy places
where the alder & ferns grow can restore my spirits —

(I made it an object to find a new parmelia caperata in fruit in each walk) At this season methinks we do not regard the larger features of the landscape—as in the spring—but are absorbed in details— Thus when the meadows were flooded I looked far over them—to the distant woods & the outlines of hills which were more distinct. I should not have so much to say of extensive water or landscapes at this season— You are a little bewildered by the variety of objects. There must be a certain meagreness of details and nakedness for wide views.

2 July 1852, *Journal* 5:174

But there is less in the morning. Every poet has trembled on the verge of science.

18 July 1852, *Journal* 5:233

No man ever makes a discovery—ever an observation of the least importance—but he is advertised of the fact by a joy that surprises him. The powers thus celebrate all discovery.

8 August 1852, *Journal* 5:291

I must walk more with free senses— It is as bad to *study* stars & clouds as flowers & stones— I must let my senses wander as my thoughts—my eyes see without looking. Carlyle said that how to observe was to look— but I say that it is rather to see— & the more you look the less you will observe— I have the habit of attention to such excess that my senses get no rest—but suffer from a constant strain. Be not preoccupied with looking. Go not to the object let it come to you.

When I have found myself ever looking down & confining my gaze to the flowers—I have thought it might be well to get into the habit of observing the clouds as a corrective— But ha! that study would be just as bad— What I need is not to look at all—but a true sauntering of the eye.

13 September 1852, *Journal* 5:343-44

The delicacy of the stratification in the white sand by the RR—where they have been getting out sand for the brickyards—the delicate stratification of this great globe like the leaves of the choicest volume just shut on a ladies table— The piled up history! I am struck by the slow & delicate process by which the globe was formed.

12 October 1852, *Journal* 5:370

Many a man—when I tell him that I have been on to a *mt* asks if I took a glass with me. No doubt, I could have seen further with a glass and particular objects more distinctly —could have counted more meeting-houses; but this has nothing to do with the peculiar beauty & grandeur of the view which an elevated position affords. It was not to see a few particular objects as if they were near at hand as I had been accustomed to see them, that I ascended the *mt* —but to see an infinite variety far & near in their relation to each other thus reduced to a single picture. The facts of science in comparison with poetry are wont to be as vulgar as looking from the *mt* with a telescope. It is a counting of meeting-houses.

20 October 1852, *Journal* 5:378

Mrs Ripley told me this Pm that Russell had decided that that green (& sometimes yellow dust) on the underside

of stones in walls was a decaying state of Lepraria chlo-
rina a lichen—the yellow another species of Lepraria.

Science suggests the value of mutual intelligence. I
have long known this dust—but as I did not know the
name of it, i.e. what others called & therefore could not
conveniently speak of it— It has suggested less to me &
I have made less use of it. I now first feel as if I had got
hold of it

15 January 1853, *Journal* 5:444

I am somewhat oppressed & saddened by the sameness
& apparent poverty of the heavens—that these irregular
& few geometrical figures which the constellations make
are no other than those seen by the Chaldaean shepherds
— I pine for a new World in the heavens as well as on the
earth— And though it is some consolation to hear of the
wilderness of stars & systems invisible to the naked eye
—yet the sky does not make that impression of variety &
wildness that even the forest does as it ought— It makes
an impression rather of simplicity & and unchangeable-
ness as of eternal laws— This being the same constella-
tion which the shepherds saw & obedien still to the
same law— It does not affect me as that unhandselled
wilderness which the forest is— I seem to see it pierced
with visual rays from a thousand observatories— It is
more the domain of science than of poetry. But it is the
stars as not known to science that I would know—the
stars which the lonely traveller knows. . . . The heavens
shall be as new at least as the world is new. This classification of the stars is old and musty—it is as if a mildew
had taken place in the heavens—as if the stars so closely
packed had heated & moulded there. . . . The heavens

commonly look as dry & meagre as our astronomies are
—mere troops as the latter are catalogues of stars— The
milky way yields no milk. A few good anecdotes is our
science—with a few imposing facts respecting distance
& size—& little or nothing about the stars as they concern man—teaching how he may survey a country or sail
a ship—& not how he may steer his life— Astrology
contained the germ of a higher truth than this— It may
happen that the stars are more significant & truly celestial to the teamster than to the astronomer— Nobody
sees the stars now—they study astronomy at the district
school—& learn that the sun is 195 millions distant &
the like—a statement which never made any impression
on me because I never walked it and which I cannot be
said to believe— But the sun shines nevertheless.
Though observatories are multiplied the heavens receive
very little attention. The naked eye may easily see farther
than the armed. It depends on who looks through it—
No superior telescope to this has been invented— In
those big ones the recoil is equal to the force of the dis-
charge. The poet's eye in a fine frenzy rolling ranges from
earth to heaven—but this the astronomer's does not often do. It does not see far beyond the dome of Greenwich
observatory. Compared with the visible phenomena of
the heavens the anecdotes of science affect me as trivial &
petty. Mans eye is the true star-finder—the comet-seeker.
As I sat looking out the window the other evening just after dark I saw the lamp of a freight train—& near by just
over the train a bright star—which looked exactly like
the former as if it belonged to a different part of the same
train— It was difficult to realize that the one was a feeble
oil lamp—the other a world.

21 January 1853, *Journal* 5:446–48

The Secretary of the Association for the Ad[vancement]. of Science — requested me as he probably has thousands of others — by a printed circular letter from Washington the other day — to fill the blanks against certain questions — among which the most important one was — what branch of science I was specially interested in — Using the term science in the most comprehensive sense possible — Now though I could state to a select few that department of human inquiry which engages me — & should be rejoiced at an opportunity to do so — I felt that it would be to make myself the laughing stock of the scientific community — to describe or attempt to describe to them that branch of science which specially interests me — in as much as they do not believe in a science which deals with the higher law. So I was obliged to speak to their condition and describe to them that poor part of me which alone they can understand. The fact is I am a mystic — a transcendentalist — & a natural philosopher to boot. Now I think — of it — I should have told them at once that I was a transcendentalist — that would have been the shortest way of telling them that they would not understand my explanations.

How absurd that though I probably stand as near to nature as any of them, and am by constitution as good an observer as most — yet a true account of my relation to nature should excite their ridicule only. If it had been the secretary of an association of which Plato or Aristotle was the President — I should not have hesitated to describe my studies at once & particularly.

5 March 1853, *Journal* 5:469-70

Organization, — how it prevails! After a little discipline, we study with love and reverence the forms of disease as healthy organisms. The fungi have a department in the science of botany. Who can doubt but that they too are fungi lower in the scale which he sees on the wick of his lamp!

15 March 1853, *Journal* V:21

Might not my Journal be called "Field Notes?"

21 March 1853, *Journal* V:32

No sap flows from the maples I cut into, except that one in Lincoln. What means it? *Hylodes Pickeringii*, a name that is longer than the frog itself! A description of animals, too, from a dead specimen only, as if, in a work on man, you were to describe a dead man only, omitting his manners and customs, his institutions and divine faculties, from want of opportunity to observe them, suggesting, perchance, that the colors of the eye are said to be much more brilliant in the living specimen, and that some cannibal, your neighbor, who has tried him on his table, has found him to be sweet and nutritious, good on the gridiron. Having had no opportunity to observe his habits, because you do not live in the country. Only dindons and dandies. Nothing is known of his habits. Food: seeds of wheat, beef, pork, and potatoes.

22 March 1853, *Journal* V:39

One studies books of science merely to learn the language of naturalists, — to be able to communicate with them. . . .

Man cannot afford to be a naturalist, to look at Nature directly, but only with the side of his eye. He must look through and beyond her. To look at her is fatal as to look at the head of Medusa. It turns the man of science to

stone. I feel that I am dissipated by so many observations. I should be the magnet in the midst of all this dust
and filings. I knock the back of my hand against a rock,
and as I smooth back the skin, I find myself prepared to
study lichens there. I look upon man but as a fungus. I
have almost a slight, dry headache as the result of all this
observing. How to observe is how to behave. O for a little Lethe! To crown all, lichens, which are so thin, are described in the *dry* state, as they are most commonly, not
most truly, seen. Truly, they are *dryly* described.

23 March 1853, *Journal* V:42–45

When I saw the fungi in my lamp, I was startled and
awed, as if I were stooping too low, and should next be
found classifying carbuncles and ulcers. Is there not
sense in the mass of men who ignore and confound these
things, and never see the cryptogamia on the one side any
more than the stars on the other? Underfoot they catch a
transient glimpse of what they call toadstools, mosses,
and frog-spittle, and overhead of the heavens, but they
can all read the pillars on a Mexican quarter. They ignore
the worlds above and below, keep straight along, and do
not run their boots down at the heel as I do.

25 March 1853, *Journal* V:51

Ah, those youthful days! are they never to return? when
the walker does not too curiously observe particulars, but
sees, hears, scents, tastes, and feels only himself, — the
phenomena that show themselves in him, — his expand-
ing body, his intellect and heart. No worm or insect,
quadruped or bird, confined his view, but the un-

bounded universe was his. A bird is now become a mote in his eye.

30 March 1853, *Journal* V:75

If you make the least correct observation of nature this year, you will have occasion to repeat it with illustrations the next, and the season and life itself is prolonged.

7 April 1853, *Journal* V:100

He is the richest who has most use for nature as raw material of tropes and symbols with which to describe his life. If these gates of golden willows affect me, they correspond to the beauty and promise of some experience on which I am entering. If I am overflowing with life, am rich in experience for which I lack expression, then nature will be my language full of poetry,—all nature will *fable*, and every natural phenomenon be a myth. The man of science, who is not seeking for expression but for a fact to be expressed merely, studies nature as a dead language. I pray for such inward experience as will make nature significant.

10 May 1853, *Journal* V:135

I quarrel with most botanists' description of different species, say of willows. It is a difference without a distinction. No stress is laid upon the peculiarity of the species in question, and it requires a very careful examination and comparison to detect any difference in the description. Having described you one species, he begins again at the beginning when he comes to the next and describes it *absolutely*, wasting time; in fact does not describe the species, but rather the genus or family; as if, in describing the particular races of men, you should say of each in its turn that it is but dust and to dust it shall return. The object should be to describe not those particulars in which a species resembles its genus, for they are many and that would be but a negative description, but those in which it is peculiar, for they are few and positive.

25 May 1853, *Journal* V:188–89

Is it not as language that all natural objects affect the poet? He sees a flower or other object, and it is beautiful or affecting to him because it is a symbol of his thought, and what he indistinctly feels or perceives is matured in some other organization. The objects I behold correspond to my mood.

7 August 1853, *Journal* V:359

Conchologists call those shells "which are fished up from the depths of the ocean" and are never seen on the shore, which are the rarest and most beautiful, *Pelagii*, but those which are cast on shore and are never so delicate and beautiful as the former, on account of exposure and abrasion, *Littorales*. So it is with the thoughts of poets: some are fresh from the deep sea, radiant with imagined beauty, — *Pelagii*; but others are comparatively worn, having been tossed by many a tide, — *Littorales*, — scaled off, abraded, and eaten by worms.

18 November 1853, *Journal* V:509

The cocks are the only birds I hear, but they are a host. They crow as freshly and bravely as ever, while poets go down the stream, degenerate into science and prose.

23 November 1853, *Journal* V:516

Saw at the Natural History rooms the skeleton of a moose with horns. The length of the spinal processes (?) over the shoulder was very great. The hind legs were longer than the front, and the horns rose about two feet above the shoulders and spread between four and five, I judged.

28 November 1853, *Journal* V:521

The skeleton which at first sight excites only a shudder in all mortals becomes at last not only a pure but suggestive and pleasing object to science. The more we know of it, the less we associate it with any goblin of our imaginations. The longer we keep it, the less likely it is that any such will come to claim it. We discover that the only spirit which haunts it is a universal intelligence which has created it in harmony with all nature. Science never saw a ghost, nor does it look for any, but it sees everywhere the traces, and it is itself the agent, of a Universal Intelligence.

2 December 1853, *Journal* VI:4

"I am compelled to decline the membership"

Concord Dec. 19th 1853

Spencer F. Baird,
Dear Sir,

I wish hereby to convey my thanks to the one who so kindly proposed me as a member of the Association for the Advancement of Science, and also to express my interest in the Association itself. Nevertheless, for the same reason that I should not be able to attend the meetings, unless held in my immediate vicinity, I am compelled to decline the membership.

> Yrs, with hearty thanks,
> Henry D. Thoreau

[Questionnaire]

Name   Henry D(avid) Thoreau

Occupation (Professional, or otherwise).   Literary and Scientific, combined with Land-surveying

Post-office address   Henry D. Thoreau Concord Mass.

Branches of science in which especial interest is felt   The Manners & Customs of the Indians of the Algonquin Group previous to contact with the civilized man.

Remarks   I may add that I am an observer of nature generally, and the character of my observations, so far as they are scientific, may be inferred from the fact that I am especially attracted by such books of science as White's Selborne and Humboldt's "Aspects of Nature."

With thanks for your "Directions," received long since I remain

> Yrs &c
> Henry D. Thoreau.
> *Correspondence,* 309–10

That sand foliage! It convinces me that Nature is still in her youth,—that florid fact about which mythology merely mutters,—that the very soil can fabulate as well as you or I. It stretches forth its baby fingers on every side. Fresh curls spring forth from its bald brow. There is nothing inorganic. This earth is not, then, a mere fragment of dead history, strata upon strata, like the leaves of a book, an object for a museum and an antiquarian, but living poetry, like the leaves of a tree, — not a fossil earth, but a living specimen.

5 February 1854, Journal VI:99–100

Much study a weariness of the flesh, eh? But did not they intend that we should read and ponder, who covered the whole earth with alphabets, — primers or bibles, — coarse or fine print? The very débris of the cliffs — the stivers[?] of the rocks — are covered with geographic lichens: no surface is permitted to be bare long. As by an inevitable decree, we have come to times at last when our very waste paper is printed. Was not He who creates lichens the abettor of Cadmus when he invented letters? Types almost arrange themselves into words and sentences as dust arranges itself under the magnet. Print! it is a close-hugging lichen that forms on a favorable surface, which paper offers. The linen gets itself wrought into paper that the song of the shirt may be printed on it. Who placed us with eyes between a microscopic and a telescopic world?

19 February 1854, Journal VI:132–33

The sand foliage is vital in its form, reminding me [of] what are called the vitals of the animal body. I am not sure that its arteries are ever hollow. They are rather meandering channels with remarkably distinct sharp edges, formed instantaneously as by magic. How rapidly and perfectly it organizes itself! The material must be sufficiently cohesive. I suspect that a certain portion of clay is necessary. Mixed sand and clay being saturated with melted ice and snow, the most liquid portion flows downward through the mass, forming for itself instantly a perfect canal, using the best materials the mass affords for its banks. It digs and builds it in a twinkling. The less fluid portions clog the artery, change its course, and form thick stems and leaves. The lobe principle, — lobe of the ear (*labor, lapsus?*).

On the outside all the life of the earth is expressed in the animal or vegetable, but make a deep cut in it and you find it vital; you find in the very sands an anticipation of the vegetable leaf. No wonder, then, that plants grow and spring in it. The atoms have already learned the law. Let a vegetable sap convey it upwards and you have a vegetable leaf. No wonder that the earth expresses itself outwardly in leaves, which labors with the idea thus inwardly. The overhanging leaf sees here its prototype. The earth is pregnant with law.

2 March 1854, *Journal* VI:148

Bought a telescope to-day for eight dollars. Best military spyglass with six slides, which shuts up to about same size, fifteen dollars, and very powerful. Saw the squares of achromatic glass from Paris which Clark(e?) uses; fifty-odd dollars apiece, the larger. It takes two together, one called the flint. These French glasses all one quality of glass. My glass tried by Clark and approved.

13 March 1854, *Journal* VI:166–67

Counted over forty robins with my glass in the meadow north of Sleepy Hollow, in the grass and on the snow.

14 March 1854, *Journal* VI:167–68

At sunrise to Clamshell Hill.

River skimmed over at Willow Bay last night. Thought I should find ducks cornered up by the ice; they get behind this hill for shelter. Saw what looked like clods of plowed meadow rising above the ice. Looked with glass and found it to be more than thirty black ducks asleep with their heads in [*sic*] their backs, motionless, and thin ice formed about them. Soon one or two were moving about slowly. There was an open space, eight or ten rods by one or two. At first all within a space of apparently less than a rod [in] diameter. It was 6:30 a.m., and the sun shining on them, but bitter cold. How tough they are! I crawled far on my stomach and got a near view of them, thirty rods off. At length they detected me and quacked.

21 March 1854, *Journal* VI:173–74

Read an interesting article on Étienne Geoffroy Saint-Hilaire, the friend and contemporary of Cuvier, though opposed to him in his philosophy. He believed species to be variable. In looking for anatomical resemblances he found that he could not safely be guided by function, form, structure, size, color, etc., but only by the relative position and mutual dependence of organs. Hence his *Le Principe des Connexions* and his maxims, "An organ is sooner destroyed than transposed." . . . A principal formula of his was, "Unity of Plan, Unity of Composition."

30 March 1854, *Journal* VI:178–79

I bought me a spy-glass some weeks since. I buy but few things, and those not till long after I begin to want them, so that when I do get them I am prepared to make a perfect use of them and extract their whole sweet.

10 April 1854, *Journal* VI:192

It is remarkable how the American mind runs to statistics. Consider the number of meteorological observers and other annual phenomena. The Smithsonian Institution is a truly national institution. Every shopkeeper makes a record of the arrival of the first martin or bluebird to his box. Dodd, the broker, told me last spring that he knew when the first bluebird came to his boxes, he made a memorandum of it: John Brown, merchant, tells me this morning that the martins first came to his box on the 13th, he "made a minute of it." Beside so many entries in their day-books and ledgers, they record these things.

17 April 1854, *Journal* VI:200

Do I ever see the marsh hawk? Is it not the sharp-shinned which I have mistaken for it? A man came to me yesterday to offer me as a naturalist a two-headed calf which his cow had brought forth, but I felt nothing but disgust at the idea and began to ask myself what enormity I had committed to have such an offer made to me. I am not interested in mere phenomena, though it were the explosion of a planet, only as it may have lain in the experience of a human being.

19 April 1854, *Journal* VI:206

There is no such thing as pure *objective* observation. Your observation, to be interesting, *i.e.* to be significant, must be *subjective*. The sum of what the writer of whatever class has to report is simply some human experience, whether he be poet or philosopher or man of science. The man of most science is the man most alive, whose life is the greatest event. Senses that take cognizance of outward things merely are of no avail. It matters not where or how far you travel, — the farther commonly the worse, — but how much alive you are. If it is possible to conceive of an event outside to humanity, it is not of the slightest significance, though it were the explosion of a planet. Every important worker will report what life there is in him. It makes no odds into what seeming deserts the poet is born. Though all his neighbors pronounce it a Sahara, it will be a paradise to him; for the desert which we see is the result of the barrenness of our experience. No mere willful activity whatever, whether in writing verses or collecting statistics, will produce true poetry or science. If you are really a sick man, it is indeed to be regretted, for you cannot accomplish so much as if you were well. All that a man has to say or do that can possibly concern mankind, is in some shape or other to tell the story of his love, — to sing; and, if he is fortunate and keeps alive, he will be forever in love. This alone is to be alive to the extremities. It is a pity that this divine creature should ever suffer from cold feet; a still greater pity that the coldness so often reaches to his heart. I look over the report of the doings of a scientific association and am surprised that there is so little life to be reported; I am put off with a parcel of dry technical terms. Anything living is easily and naturally expressed in popular language. I cannot help suspecting that the life of these learned professors has been almost as inhuman and wooden as a rain-gauge or self-registering magnetic machine. They communicate no fact which rises to the temperature of blood-heat. It does n't all amount to one rhyme.

6 May 1854, *Journal* VI:236–38

In Boston yesterday an ornithologist said significantly, "If you held the bird in your hand—;" but I would rather hold it in my affections.

10 May 1854, *Journal* VI:253

Have just been looking at Nuttall's "North American Sylva." Much research, fine plates and print and paper, and unobjectionable periods, but no turpentine, or balsam, or quercitron, or salicin, or birch wine, or the aroma of the balm-of-Gilead, no gallic, or ulmic, or even malic acid. The plates are greener and higher-colored than the words, etc., etc. It is sapless, if not leafless.

15 May 1854, *Journal* VI:265

We soon get through with Nature. She excites an expectation which she cannot satisfy. The merest child which has rambled into a copsewood dreams of a wilderness so wild and strange and inexhaustible as Nature can never show him. The red-bird which I saw on my companion's string on election days I thought but the outmost sentinel of the wild, immortal camp,—of the wild and dazzling infantry of the wilderness,—that the deeper woods abounded with redder birds still; but, now that I have threaded all our woods and waded the swamps, I have never yet met with his compeer, still less his wilder kindred. . . . I expected a fauna more infinite and various, birds of more dazzling colors and more celestial song. How many springs shall I continue to see the common sucker (*Catostomus Bostoniensis*) floating dead on our river! Will not Nature select her types from a new fount? The vignette of the year. This earth which is spread out like a map around me is but the lining of my inmost soul exposed. In me is the sucker that I see. No wholly extraneous object can compel me to recognize it. I am guilty of suckers.

23 May 1854, *Journal* VI:293–94

The inhumanity of science concerns me, as when I am tempted to kill a rare snake that I may ascertain its species. I feel that this is not the means of acquiring true knowledge.

28 May 1854, *Journal* VI:311

Do you not feel the fruit of your spring and summer beginning to ripen, to harden its seed within you? Do not your thoughts begin to acquire consistency as well as flavor and ripeness? How can we expect a harvest of thought who have not had a seed-time of character? Already some of my small thoughts—fruit of my spring life—are ripe, like the berries which feed the first broods of birds; and other some are prematurely ripe and bright, like the lower leaves of the herbs which have felt the summer's drought.

7 August 1854, *Journal* VI:426

"Walden" published. Elder-berries. Waxwork yellowing.

9 August 1854, *Journal* VI:429

"As I was desirous to recover the long lost bottom of Walden Pond"

As I was desirous to recover the long lost bottom of Walden Pond, I surveyed it carefully, before the ice broke up, early in '46, with compass and chain and sounding line. There have been many stories told about the bottom, or rather no bottom, of this pond, which certainly had no foundation for themselves. It is remarkable how long men will believe in the bottomlessness of a pond without taking the trouble to sound it. I have visited two such Bottomless Ponds in one walk in this neighborhood. Many have believed that Walden reached quite through to the other side of the globe. Some who have lain flat on the ice for a long time, looking down through the illusive medium, perchance with watery eyes into the bargain, and driven to hasty conclusions by the fear of catching cold in their breasts, have seen vast holes "into which a load of hay might be driven," if there were any body to drive it, the undoubted source of the Styx and entrance to the Infernal Regions from these parts. Others have gone down from the village with a "fifty-six" and a wagon load of inch rope, but yet have failed to find any bottom; for while the "fifty-six" was resting by the way, they were paying out the rope in the vain attempt to fathom their truly immeasurable capacity for marvellousness. But I can assure my readers that Walden has a reasonably tight bottom at a not unreasonable, though at an unusual, depth. I fathomed it easily with a cod-line and a stone weighing about a pound and a half, and could tell accurately when the stone left the bottom, by having to pull so much harder before the water got underneath to

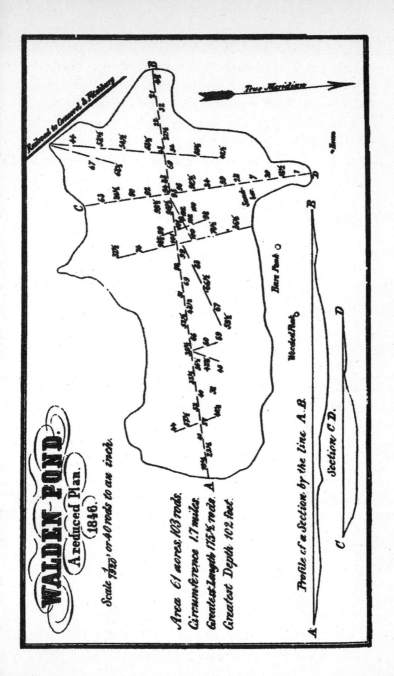

WALDEN-POND.
A reduced Plan.
(1846.)

Scale $\frac{1}{7920}$, or 40 rods to an inch.

Area 61 acres, 103 rods.
Circumference 1.7 miles.
Greatest-Length 175⅗ rods.
Greatest Depth 102 feet.

Railroad to Concord & Fitchburg

True Meridian

Bare Peak

Wooded Peak

Profile of a Section by the line A.B.

Section C.D.

A   B

C   D

help me. The greatest depth was exactly one hundred and two feet; to which may be added the five feet which it has risen since, making one hundred and seven. This is a remarkable depth for so small an area; yet not an inch of it can be spared by the imagination. What if all ponds were shallow? Would it not react on the minds of men? I am thankful that this pond was made deep and pure for a symbol. While men believe in the infinite some ponds will be thought to be bottomless.

❦

When I had mapped the pond by the scale of ten rods to an inch, and put down the soundings, more than a hundred in all, I observed this remarkable coincidence. Having noticed that the number indicating the greatest depth was apparently in the centre of the map, I laid a rule on the map lengthwise, and then breadthwise, and found, to my surprise, that the line of greatest length intersected the line of greatest breadth *exactly* at the point of greatest depth, notwithstanding that the middle is so nearly level, the outline of the pond far from regular, and the extreme length and breadth were got by measuring into the coves; and I said to myself, Who knows but this hint would conduct to the deepest part of the ocean as well as of a pond or puddle? Is not this the rule also for the height of mountains, regarded as the opposite of valleys? We know that a hill is not highest at its narrowest part.

Of five coves, three, or all which had been sounded, were observed to have a bar quite across their mouths and deeper water within, so that the bay tended to be an expansion of water within the land not only horizontally but vertically, and to form a basin or independent pond, the direction of the two capes showing the course of the bar. Every harbor on the sea-coast, also, has its bar at its entrance. In proportion as the mouth of the cove was wider compared with its length, the water over the bar was deeper compared with that in the basin. Given, then, the length and breadth of the cove, and the character of the surrounding shore, and you have almost elements enough to make out a formula for all cases.

In order to see how nearly I could guess, with this experience, at the deepest point in a pond, by observing the outlines of its surface and the character of its shores alone, I made a plan of White Pond, which contains about forty-one acres, and, like this, has no island in it, nor any visible inlet or outlet; and as the line of greatest breadth fell very near the line of least breadth, where two opposite capes approached each other and two opposite bays receded, I ventured to mark a point a short distance from the latter line, but still on the line of greatest length, as the deepest. The deepest part was found to be within one hundred feet of this, still farther in the direction to which I had inclined, and was only one foot deeper, namely, sixty feet. Of course, a stream running through, or an island in the pond, would make the problem much more complicated.

If we knew all the laws of Nature, we should need only one fact, or the description of one actual phenomenon, to infer all the particular results at that point. Now we know only a few laws, and our result is vitiated, not, of course, by any confusion or irregularity in Nature, but by our ignorance of essential elements in the calculation. Our notions of law and harmony are commonly confined to those instances which we detect; but the harmony which results from a far greater number of seemingly conflict-

ing, but really concurring, laws, which we have not detected, is still more wonderful. The particular laws are as
our points of view, as, to the traveller, a mountain outline
varies with every step, and it has an infinite number of
profiles, though absolutely but one form. Even when cleft
or bored through it is not comprehended in its entire-
ness.

What I have observed of the pond is no less true in
ethics. It is the law of average. Such a rule of the two diameters not only guides us toward the sun in the system
and the heart in man, but draw lines through the length
and breadth of the aggregate of a man's particular daily
behaviors and waves of life into his coves and inlets, and
where they intersect will be the height or depth of his
character. Perhaps we need only to know how his shores
trend and his adjacent country or circumstances, to infer
his depth and concealed bottom. If he is surrounded by
mountainous circumstances, an Achillean shore, whose
peaks overshadow and are reflected in his bosom, they
suggest a corresponding depth in him. But a low and
smooth shore proves him shallow on that side. In our
bodies, a bold projecting brow falls off to and indicates a
corresponding depth of thought. Also there is a bar
across the entrance of our every cove, or particular inclination; each is our harbor for a season, in which we are
detained and partially land-locked. These inclinations
are not whimsical usually, but their form, size, and direction are determined by the promontories of the shore, the
ancient axes of elevation. When this bar is gradually in-
creased by storms, tides, or currents, or there is a subsi-
dence of the waters, so that it reaches to the surface, that
which was at first but an inclination in the shore in which

a thought was harbored becomes an individual lake, cut off from the ocean, wherein the thought secures its own conditions, changes, perhaps, from salt to fresh, becomes a sweet sea, dead sea, or a marsh. At the advent of each individual into this life, may we not suppose that such a bar has risen to the surface somewhere? It is true, we are such poor navigators that our thoughts, for the most part, stand off and on upon a harborless coast, are conversant only with the bights of the bays of poesy, or steer for the public ports of entry, and go into the dry docks of science, where they merely refit for this world, and no natural currents concur to individualize them.

"The Pond in Winter," *Walden,* 285–87, 289–92

IV

❦

"My expectation
ripens to discovery"

August 1854–November 1858

I have just been through the process of killing the cistudo [box turtle] for the sake of science; but I cannot excuse myself for this murder, and see that such actions are inconsistent with the poetic perception, however they may serve science, and will affect the quality of my observations. I pray that I may walk more innocently and serenely through nature. No reasoning whatever reconciles me to this act. It affects my day injuriously. I have lost some self-respect. I have a murderer's experience in a degree.

18 August 1854, *Journal* VI:452

When I look at the stars, nothing which the astronomers have said attaches to them, they are so simple and remote. *Their* knowledge is felt to be all terrestrial and to concern the earth alone. It suggests that the same is the case with every object, however familiar; our so-called knowledge of it is equally vulgar and remote.

One might say that all views through a telescope or microscope were purely visionary, for it is only by his eye and not by any other sense—not by his whole man—

that the beholder is there where he is presumed to be. It is a disruptive mode of viewing as far as the beholder is concerned.

29 September 1854, Journal VII:60–61

Men have been talking now for a week at the post-office about the age of the great elm, as a matter interesting but impossible to be determined. The very choppers and travellers have stood upon its prostrate trunk and speculated upon its age, as if it were a profound mystery. I stooped and read its years to them (127 at nine and a half feet), but they heard me as the wind that once sighed through its branches. They still surmised that it might be two hundred years old, but they never stooped to read the inscription. Truly they love darkness rather than light. One said it was probably one hundred and fifty, for he had heard somebody say that for fifty years the elm grew, for fifty it stood still, and for fifty it was dying. (Wonder what portion of his career he stood still!) Truly all men are not men of science. They dwell within an integument of prejudice thicker than the bark of the cork-tree, but it is valuable chiefly to stop bottles with. Tied to their buoyant prejudices, they keep themselves afloat when honest swimmers sink.

26 January 1856, Journal VIII:145–46

Again, as so many times, I [am] reminded of the advantage to the poet, and philosopher, and naturalist, and whomsoever, of pursuing from time to time some other business than his chosen one, — seeing with the side of the eye. The poet will so get visions which no deliberate abandonment can secure. The philosopher is so forced to recognize principles which long study might not detect. And the naturalist even will stumble upon some new and unexpected flower or animal. . . .

Observing the young pitch pines by the road south of Loring's lot that was so heavily wooded, George Hubbard remarked that if they were cut down oaks would spring up, and sure enough, looking across the road to where Loring's white pines recently stood so densely, the ground was all covered with young oaks. *Mem.* — Let me look at the site of some thick pine woods which I remember, and see what has sprung up; *e.g.* the pitch pines on Thrush Alley and the white pines on Cliffs, also at Baker's chestnuts, and the chestnut lot on the Tim. Brooks farm.

28 April 1856, *Journal* VIII:314–16

I suspect that I can throw a little light on the fact that when a dense pine wood is cut down oaks, etc., may take its place. There were only pines, no other tree. They are cut off, and, after two years have elapsed, you see oaks, or perhaps a few other hard woods, springing up with scarcely a pine amid them, and you wonder how the acorns could have lain in the ground so long without decaying. There is a good example at Loring's lot. But if you look through a thick pine wood, even the exclusively pitch pine ones, you will detect many little oaks, birches, etc., sprung probably from seeds carried into the thicket by squirrels, etc., and blown thither, but which are overshadowed and choked by the pines. This planting under the shelter of the pines may be carried on annually, and the plants annually die, but when the pines are cleared off, the oaks, etc., having got just the start they want, and now secured favorable conditions, immediately spring up to trees. Scarcely enough allowance has been made for the agency of squirrels and birds in dispersing seeds.

13 May 1856, *Journal* VIII:335

While drinking at Assabet Spring in woods, noticed a cherry-stone on the bottom. A bird that came to drink must have brought it half a mile. So the tree gets planted!

13 July 1856, *Journal* VIII:411

My plants in press are in a sad condition; mildew has invaded them during the late damp weather, even those that were nearly dry. I find more and other plants than I counted on. Very bad weather of late for pressing plants. Give me the dry heat of July. Even growing leaves out of doors are spotted with fungi now, much more than mine in press.

16 August 1856, *Journal* IX:6

I see that all is not garden and cultivated field and crops, that there are square rods in Middlesex County as purely primitive and wild as they were a thousand years ago, which have escaped the plow and the axe and the scythe and the cranberry-rake, little oases of wildness in the desert of our civilization, wild as a square rod on the moon, supposing it to be uninhabited. I believe almost in the personality of such planetary matter, feel something akin to reverence for it, can even worship it as terrene, titanic matter extant in my day. We are so different we admire each other, we healthily attract one another. I love it as a maiden. These spots are meteoric, aerolitic, and such matter has in all ages been worshipped. Aye, when we are lifted out of the slime and film of our habitual life, we see the whole globe to be an aerolite, and reverence it as such, and make pilgrimages to it, far off as it is. How happens it that we reverence the stones which fall from another planet, and not the stones which belong to this,—another globe, not this,—heaven, and not earth? Are not the stones in Hodge's wall as good as the aerolite at Mecca? Is not our broad back-door-stone as good as any corner-stone in heaven?

It would imply the regeneration of mankind, if they were to become elevated enough to truly worship stocks and stones.

30 August 1856, *Journal* IX:44–45

I think we may detect that some sort of preparation and faint expectation preceded every discovery we have made. We blunder into no discovery but it will appear that we have prayed and disciplined ourselves for it. Some years ago I sought for Indian hemp (*Apocynum cannabinum*) hereabouts in vain, and concluded that it did not grow here. A month or two ago I read again, as many times before, that its blossoms were very small, scarcely a third as large as those of the common species, and for some unaccountable reason this distinction kept recurring to me, and I regarded the size of the flowers I saw, though I did not believe that it grew here; and in a day or two my eyes fell on [it], aye, in three different places, and different varieties of it. . . .

It commonly chances that I make my most interesting botanical discoveries when I [am] in a thrilled and expectant mood, perhaps wading in some remote swamp where I have just found something novel and feel more than usually remote from the town. Or some rare plant which for some reason has occupied a strangely promi-

nent place in my thoughts for some time will present itself. My expectation ripens to discovery. I am prepared
for strange things.

2 September 1856, *Journal* IX:53–54

My first botany, as I remember, was Bigelow's "Plants of
Boston and Vicinity," which I began to use about twenty
years ago, looking chiefly for the popular names and the
short references to the localities of plants, even without
any regard to the plant. I also learned the names of many,
but without using any system, and forgot them soon. I
was not inclined to pluck flowers; preferred to leave them
where they were, liked them best there. I was never in the
least interested in plants in the house. But from year to
year we look at Nature with new eyes. About half a dozen
years ago I found myself again attending to plants with
more method, looking out the name of each one and remembering it. I began to bring them home in my hat, a
straw one with a scaffold lining to it, which I called my
botany-box. I never used any other, and when some
whom I visited were evidently surprised at its dilapidated
look, as I deposited it on their front entry table, I assured
them it was not so much my hat as my botany-box. I remember gazing with interest at the swamps about those
days and wondering if I could ever attain to such famil-

iarity with plants that I should know the species of every twig and leaf in them, that I should be acquainted with every plant (excepting grasses and cryptogamous ones), summer and winter, that I saw. Though I knew most of the flowers, and there were not in any particular swamp more than half a dozen shrubs that I did not know, yet these made it seem like a maze to me, of a thousand strange species, and I even thought of commencing at one end and looking it faithfully and laboriously through till I knew it all. I little thought that in a year or two I should have attained to that knowledge without all that labor. Still I never studied botany, and do not to-day systematically, the most natural system is still so artificial. I wanted to know my neighbors, if possible, — to get a little nearer to them. I soon found myself observing when plants first blossomed and leafed, and I followed it up early and late, far and near, several years in succession, running to different sides of the town and into the neighboring towns, often between twenty and thirty miles in a day. I often visited a particular plant four or five miles distant, half a dozen times within a fortnight, that I might know exactly when it opened, beside attending to a great many others in different directions and some of them equally distant, at the same time. At the same time I had an eye for birds and whatever else might offer.

4 December 1856, *Journal* IX:156–58

The most poetic and truest account of objects is generally by those who first observe them, or the discoverers of them, whether a sharper perception and curiosity in them led to the discovery or the greater novelty more inspired their report. Accordingly I love most to read the

accounts of a country, its natural productions and curiosities, by those who first settled it, and also the earliest,
though often unscientific, writers on natural science.

27 January 1857, *Journal* IX:232

What is the relation between a bird and the ear that ap-
preciates its melody, to whom, perchance, it is more
charming and significant than to any else? Certainly they
are intimately related, and the one was made for the
other. It is a natural fact. If I were to discover that a certain kind of stone by the pond-shore was affected, say
partially disintegrated, by a particular natural sound, as
of a bird or insect, I see that one could not be completely
described without describing the other. I am that rock by
the pond-side.

What is hope, what is expectation, but a seed-time
whose harvest cannot fail, an irresistible expedition of
the mind, at length to be victorious?

20 February 1857, *Journal* IX:274–75

Dine with Agassiz at R. W. E.'s. He thinks that the suckers die of asphyxia, having very large air-bladders and being in the habit of coming to the surface for air. But then,
he is thinking of a different phenomenon from the one I
speak of, which last is confined to the very earliest spring
or winter. He says that the *Emys picta* does not copulate
till seven years old, and then does not lay till four years after copulation, or when eleven years old. The *Cistuda
Blandingii* (which he has heard of in Massachusetts only
at Lancaster) copulates at eight or nine years of age. He
says this is not a *Cistuda* but an *Emys*. He has eggs of the
serpentina from which the young did not come forth till

the next spring. He thinks that the Esquimau dog is the only indigenous one in the United States. He had not observed the silvery appearance and the dryness of the lycoperdon fungus in water which I showed. He had broken caterpillars and found the crystals of ice in them, but had not thawed them. When I began to tell him of my experiment on a frozen fish, he said that Pallas had shown that fishes were frozen and thawed again, but I affirmed the contrary, and then Agassiz agreed with me.

20 March 1857, *Journal* IX:298–99

I saw a red squirrel run along the bank under the hemlocks with a nut in its mouth. He stopped near the foot of a hemlock, and, hastily pawing a hole with his fore feet, dropped the nut, covered it up, and retreated part way up the trunk of the tree, all in a few moments. I approached the shore to examine the deposit, and he, descending betrayed no little anxiety for his treasure and made two or three motions to recover the nut before he retreated. Digging there, I found two pignuts joined together, with their green shells on, buried about an inch and a half in the soil, under the red hemlock leaves. This, then, is the way forests are planted. This nut must have been brought twenty rods at least and was buried at just the right depth. If the squirrel is killed, or neglects its deposit, a hickory springs up.

P.M. — I walked to that very dense and handsome white pine grove east of Beck Stow's Swamp. It is about fifteen rods square, the trees large, ten to twenty inches in diameter. It is separated by a wall from another pine wood with a few oaks in it on the southeast, and about thirty rods north and west are other pine and oak woods.

Standing on the edge of the wood and looking through it, —for it is quite level and free from underwood, *mostly* bare, red-carpeted ground,—you would have said that there was not a hardwood tree in it, young or old, though I afterward found on one edge a middling-sized sassafras, a birch, a small tupelo, and two little scarlet oaks, but, what was more interesting, I found, on looking closely over its floor, that, alternating with thin ferns and small blueberry bushes, there was, as often as every five feet, a little oak, three to twelve inches high, and in one place I found a green acorn dropped by the base of a tree. I was surprised, I confess, to find my own theory so perfectly proved.

24 September 1857, *Journal* X:39–41

Sometimes I would rather get a transient glimpse or side view of a thing than stand fronting to it,—as those poly-podies. The object I caught a glimpse of as I went by haunts my thoughts a long time, is infinitely suggestive, and I do not care to front it and scrutinize it, for I know that the thing that really concerns me is not there, but in my relation to that. That is a mere reflecting surface. . . .

I think that the man of science makes this mistake, and the mass of mankind along with him: that you should coolly give your chief attention to the phenomenon which excites you as something independent on you, and not as it is related to you. The important fact is its effect on me. He thinks that I have no business to see anything else but just what he defines the rainbow to be, but I care not whether my vision of truth is a waking thought or dream remembered, whether it is seen in the light or in the dark. It is the subject of the vision, the truth alone, that concerns me. The philosopher for whom rainbows, etc., can be explained away never saw them. With regard to such objects, I find that it is not they themselves (with which the men of science deal) that concern me; the point of interest is somewhere *between* me and them (*i.e.* the objects).

5 November 1857, *Journal* X:164–65

We read the English poets; we study botany and zoölogy and geology, lean and dry as they are; and it is rare that we get a new suggestion. It is ebb-tide with the scientific reports, Professor —— in the chair. We would fain know something more about these animals and stones and trees around us. We are ready to skin the animals alive to come at them. Our scientific names convey a very partial information only; they suggest certain thoughts only. It does not occur to me that there are other names for most of these objects, given by a people who stood between me and them, who had better senses than our race. How little I know of that *arbor-vitae* when I have learned only what science can tell me! It is but a word. It is not a *tree* of *life*. But there are twenty words for the tree and its different parts which the Indian gave, which are not in our botanies, which imply a more practical and vital science. He used it every day. He was well acquainted with its wood, and its bark, and its leaves. No science does more than arrange what knowledge we have of any class of objects. But, generally speaking, how much more conversant was the Indian with any wild animal or plant than we are, and in his language is implied all that intimacy, as much as ours is expressed in our language. How many words in his language about a moose, or birch bark, and the like! The Indian stood nearer to wild nature than we. The wildest and noblest quadrupeds, even the largest fresh-water fishes, some of the wildest and noblest birds and the fairest flowers have actually receded as *we* advanced, and we have but the most distant knowledge of them. A rumor has come down to us that the skin of a lion was seen and his roar heard here by an early settler. But there was a race here that slept on his skin. It was a new light when my guide gave me Indian names for things for which I had only scientific ones before. In proportion as I understood the language, I saw them from a new point of view.

5 March 1858, *Journal* X:293–95

In a shallow rain-water pool [on Mt. Monadnock], or rock cistern, about three rods long by one or one and a half wide, several hundred feet below the summit, on the west side, but still on the bare rocky top and on the steepest side of the summit, I saw toad-spawn (black with white bellies), also some very large spawn new to me. There were four or five masses of it, each three or four inches in diameter and of a peculiar light misty bluish white as it lay in the water near the surface, attached to some weed or stick, as usual. . . . This pool was bounded on one or two sides by those rounded walls of rock five or six feet high. My companion had said that he heard a bullfrog the evening before. Is it likely that these toads and frogs ever hopped up there? The hylodes peeped regularly toward night each day in a similar pool much nearer the summit. Agassiz might say that they originated on the top. Perhaps they fell from the clouds in the form of spawn or tadpoles or young frogs. I think it more likely that they fell down than that they hopped up. Yet how can they escape the frosts of winter? The mud is hardly deep enough to protect them.

3 June 1858, *Journal* X:467–68

Emerson says that he and Agassiz and Company broke some dozens of ale-bottles, one after another, with their bullets, in the Adirondack country, using them for marks! It sounds rather Cockneyish. He says that he shot a peetweet for Agassiz, and this, I think he said, was the first game he ever bagged. He carried a double-barrelled gun, — rifle and shotgun, — which he bought for the purpose, which he says received much commendation, — all parties thought it a very pretty piece. Think of Emerson shooting a peetweet (with shot) for Agassiz, and cracking an ale-bottle (after emptying it) with his rifle at six rods! They cut several pounds of lead out of the tree. It is just what Mike Saunders, the merchant's clerk, did when he was there.

23 August 1858, *Journal* XI:119–20

How hard one must work in order to acquire his language, — words by which to express himself! I have known a particular rush, for instance, for at least twenty years, but have ever been prevented from describing some [of] its peculiarities, because I did not know its name nor any one in the neighborhood who could tell me it. With the knowledge of the name comes a distincter recognition and knowledge of the thing. That shore is now more describable, and poetic even. My knowledge was cramped and confined before, and grew rusty because not used, — for it could not be used. My knowledge now becomes communicable and grows by communication. I can now learn what others know about the same thing.

29 August 1858, *Journal* XI:137

It requires a different intention of the eye in the same locality to see different plants, as, for example, *Juncaceae* and *Gramineae* even; *i.e.*, I find that when I am looking for the former, I do not see the latter in their midst. How much more, then, it requires different intentions of the eye and of the mind to attend to different departments of knowledge! How differently the poet and the naturalist look at objects! A man sees only what concerns him. A botanist absorbed in the pursuit of grasses does not distinguish the grandest pasture oaks. He as it were tramples down oaks unwittingly in his walk.

8 September 1858, *Journal* XI:153

On our way, near the Hosmer moraine, let off some pasture thistle-down. One steadily rose from my hand, freighted with its seed, till it was several hundred feet high, and then passed out of sight eastward. Its down was particularly spreading or open. Is not here a hint to balloonists? Astronomers can calculate the orbit of that this-

tle-down called the comet, now in the northwest sky, conveying its nucleus, which may not be so solid as a thistle's seed, somewhither, but what astronomer can calculate the orbit of my thistle-down and tell where it will deposit its precious freight at last? It may still be travelling when I am sleeping.

29 September 1858, *Journal* XI:185–86

The actual objects which one person will see from a particular hilltop are just as different from those which another will see as the persons are different. The scarlet oak must, in a sense, be in your eye when you go forth. We cannot see anything until we are possessed with the idea of it, and then we can hardly see anything else. In my botanical rambles I find that first the idea, or image, of a plant occupies my thoughts, though it may at first seem very foreign to this locality, and for some weeks or months I go thinking of it and expecting it unconsciously, and at length I surely see it, and it is henceforth an actual neighbor of mine. This is the history of my finding a score or more of rare plants which I could name. . . .

Why, it takes a sharpshooter to bring down even such trivial game as snipes and woodcocks; he must take very particular aim, and know what he is aiming at. He would stand a very small chance if he fired at random into the sky, being told that snipes were flying there. And so it is with him that shoots at beauty. Not till the sky falls will he catch larks, unless he is a trained sportsman. He will not bag any if he does not already know its seasons and haunts and the color of its wing, — if he has not dreamed of it, so that he can *anticipate* it; then, indeed, he flushes it at every step, shoots double and on the wing, with both barrels, even in corn-fields.

4 November 1858, Journal XI:285–86

It is remarkable how little any but a lichenist will observe on the bark of trees. The mass of men have but the vaguest and most indefinite notion of mosses, as a sort of shreds and fringes, and the world in which the lichenist dwells is much further from theirs than one side of this earth from the other. They see bark as if they saw it not. These objects which, though constantly visible, are rarely looked at are a sort of eye-brush.

Each phase of nature, while not invisible, is yet not too distinct and obtrusive. It is there to be found when we look for it, but not demanding our attention. It is like a silent but sympathizing companion in whose company we retain most of the advantages of solitude, with whom we can walk and talk, or be silent, naturally, without the necessity of talking in a strain foreign to the place. . . .

Lichens as they affect the scenery, as picturesque objects described by Gilpin or others, are one thing; as they concern the lichenist, quite another.

8 November 1858, Journal XI:296–97

I cannot but see still in my mind's eye those little striped breams poised in Walden's glaucous water. They balance all the rest of the world in my estimation at present, for this is the bream that I have just found, and for the time I neglect all its brethren and am ready to kill the fatted calf on its account. For more than two centuries have men fished here and have not distinguished this permanent settler of the township. It is not like a new bird, a transient visitor that may not be seen again for years, but there it dwells and has dwelt permanently, who can tell how long? When my eyes first rested on Walden the striped bream was poised in it, though I did not see it, and when Tahatawan paddled his canoe there. How wild it makes the pond and the township to find a new fish in it! America renews her youth here. But in my account of this bream I cannot go a hair's breadth beyond the mere statement that it exists, — the miracle of its existence, my contemporary and neighbor, yet so different from me! I can only poise my thought there by its side and try to think like a bream for a moment. I can only think of precious jewels, of music, poetry, beauty, and the mystery of life. I only see the bream in its orbit, as I see a star, but I care not to measure its distance or weight. The bream, appreciated, floats in the pond as the centre of the system, another image of God. Its life no man can explain more than he can his own. I want you to perceive the mystery of the bream. I have a contemporary in Walden. It has fins where I have legs and arms. I have a friend among the fishes, at least a new acquaintance. Its character will interest me, I trust, not its clothes and anatomy. I do not want it to eat. Acquaintance with it is to make my life more rich and eventful. It is as if a poet or an anchorite had moved into the town, whom I can see from time to time and think of yet oftener. Perhaps there are a thousand of these striped bream which no one had thought of in that pond,—not their mere impressions in stone, but in the full tide of the bream life.

Though science may sometimes compare herself to a child picking up pebbles on the seashore, that is a rare mood with her; ordinarily her practical belief is that it is only a few pebbles which are *not* known, weighed and measured. A new species of fish signifies hardly more than a new name. See what is contributed in the scientific reports. One counts the fin-rays, another measures the intestines, a third daguerreotypes a scale, etc., etc.; otherwise there's nothing to be said. As if all but this were done, and these were very rich and generous contributions to science. Her votaries may be seen wandering along the shore of the ocean of truth, with their backs to that ocean, ready to seize on the shells which are cast up. You would say that the scientific bodies were terribly put to it for objects and subjects. A dead specimen of an animal, if it is only well preserved in alcohol, is just as good for science as a living one preserved in its native element.

What is the amount of my discovery to me? It is not that I have got one in a bottle, that it has got a name in a book, but that I have a little fishy friend in the pond. How was it when the youth first discovered fishes? Was it the number of their fin-rays or their arrangement, or the place of the fish in some system that made the boy dream of them? Is it these things that interest mankind in the fish, the inhabitant of the water? No, but a faint recognition of a living contemporary, a provoking mystery. One boy thinks of fishes and goes a-fishing from the same motive that his brother searches the poets for rare lines. It is the poetry of fishes which is their chief use; their flesh is their lowest use. The beauty of the fish, that is what it is best worth the while to measure. Its place in our systems is of comparatively little importance. Generally the boy loses some of his perception and his interest in the fish; he degenerates into a fisherman or an ichthyologist.

30 November 1858, *Journal* XI:358–60

V

*

"A fact... must be the vehicle of some humanity"

February 1859–March 1861

A good book is not made in the cheap and offhand manner of many of our scientific Reports, ushered in by the message of the President communicating it to Congress, and the order of Congress that so many thousand copies be printed, with the letters of instruction for the Secretary of the Interior (or rather exterior); the bulk of the book being a journal of a picnic or sporting expedition by a brevet Lieutenant-Colonel, illustrated by photographs of the traveller's footsteps across the plains and an admirable engraving of his native village as it appeared on leaving it, and followed by an appendix on the palaeontology of the route by a distinguished savant who was not there, the last illustrated by very finely executed engravings of some old broken shells picked up on the road.

25 February 1859, *Journal* XI:456

The mystery of the life of plants is kindred with that of our own lives, and the physiologist must not presume to explain their growth according to mechanical laws, or as he might explain some machinery of his own making. We must not expect to probe with our fingers the sanctuary of any life, whether animal or vegetable. If we do, we shall discover nothing but surface still. The ultimate expression or fruit of any created thing is a fine effluence which only the most ingenuous worshipper perceives at a reverent distance from its surface even. The cause and the effect are equally evanescent and intangible, and the former must be investigated in the same spirit and with the same reverence with which the latter is perceived. Science is often like the grub which, though it may have nestled in the germ of a fruit, has merely blighted or consumed it and never truly tasted it. Only that intellect makes any progress toward conceiving of the essence which at the same time perceives the effluence.

7 March 1859, *Journal* XII:23–24

The catechism says that the chief end of man is to glorify God and enjoy him forever, which of course is applicable mainly to God as seen in his works. Yet the only account of its beautiful insects—butterflies, etc.—which God has made and set before us which the State ever thinks of spending any money on is the account of those which are injurious to vegetation! This is the way we glorify God and enjoy him forever. Come out here and behold a thousand painted butterflies and other beautiful insects which people the air, then go to the libraries and see what kind of prayer and glorification of God is there recorded.

Massachusetts has published her report on "Insects Injurious to Vegetation," and our neighbor the "Noxious Insects of New York." We have attended to the evil and said nothing about the good. This is looking a gift horse in the mouth with a vengeance. Children are attracted by the beauty of butterflies, but their parents and legislators deem it an idle pursuit. The parents remind me of the devil, but the children of God. Though God may have pronounced his work good, we ask, "Is it not poisonous?"

Science is inhuman. Things seen with a microscope begin to be insignificant. So described, they are as monstrous as if they should be magnified a thousand diameters. Suppose I should see and describe men and houses and trees and birds as if they were a thousand times larger than they are! With our prying instruments we disturb the balance and harmony of nature.

1 May 1859, Journal XII:170–71

It is only when we forget all our learning that we begin to know. I do not get nearer by a hair's breadth to any natural object so long as I presume that I have an introduction to it from some learned man. To conceive of it with a total apprehension I must for the thousandth time approach it as something totally strange. If you would make acquaintance with the ferns you must forget your botany. You must get rid of what is commonly called *knowledge* of them. Not a single scientific term or distinction is the least to the purpose, for you would fain perceive something, and you must approach the object totally unprejudiced. You must be aware that *no thing* is what you have taken it to be. In what book is this world and its beauty described? Who has plotted the steps toward the discovery of beauty? You have got to be in a different state from common. Your greatest success will be simply to perceive that such things are, and you will have no communication to make to the Royal Society.

4 October 1859, *Journal* XII:371

A.M. — To Cambridge, where I read in Gerard's Herbal. His admirable though quaint descriptions are, to my mind, greatly superior to the modern more scientific ones. He describes not according to rule but to his natural delight in the plants. He brings them vividly before you, as one who has seen and delighted in them. It is almost as good as to see the plants themselves. It suggests that we cannot too often get rid of the barren assumption that is in our science. His leaves are leaves; his flowers, flowers; his fruit, fruit. They are green and colored and fragrant. It is a man's knowledge added to a child's delight. Modern botanical descriptions approach ever nearer to the dryness of an algebraic formula, as if $x + y$ were = to a love-letter. It is the keen joy and discrimination of the child who has just seen a flower for the first time and comes running in with it to its friends. How much better to describe your object in fresh English words rather than in these conventional Latinisms! He has really seen, and smelt, and tasted, and reports his sensations.

16 December 1859, *Journal* XIII:29–30

Aristotle, being almost if not quite the first to write systematically on animals, gives them, of course, only popular names, such as the hunters, fowlers, fishers, and farmers of his day used. He used no scientific terms. But he, having the priority and having, as it were, created science and given it its laws, those popular Greek names, even when the animal to which they were applied cannot be identified, have been in great part preserved and make those learned far-fetched and commonly unintelligible names of genera to-day, *e.g.* Ὁλοθούριον, etc., etc. His History of Animals has thus become a very storehouse of scientific nomenclature.

26 December 1859, *Journal* XIII:55

A man receives only what he is ready to receive, whether physically or intellectually or morally, as animals conceive at certain seasons their kind only. We hear and apprehend only what we already half know. If there is something which does not concern me, which is out of my line, which by experience or by genius my attention is not drawn to, however novel and remarkable it may be, if it is spoken, we hear it not, if it is written, we read it not, or if we read it, it does not detain us. Every man thus *tracks himself* through life, in all his hearing and reading and observation and travelling. His observations make a chain. The phenomenon or fact that cannot in any wise be linked with the rest which he has observed, he does not observe. By and by we may be ready to receive what we cannot receive now. I find, for example, in Aristotle something about the spawning, etc., of the pout and perch, because I know something about it already and have my attention aroused; but I do not discover till very late that he has made other equally important observations on the spawning of other fishes, because I am not interested in those fishes.

5 January 1860, *Journal* XIII:77–78

The hunter may be said to invent his game, as Neptune did the horse, and Ceres corn. . . .

Whatever aid is to be derived from the use of a scientific term, we can never begin to see anything as it is so long as we remember the scientific term which always our ignorance has imposed on it. Natural objects and phenomena are in this sense forever wild and unnamed by us.

Thus the sky and the earth sympathize, and are subject to the same laws, and in the horizon they, as it were, meet and are seen to be one.

12 February 1860, *Journal* XIII:140–41

I think that the most important requisite in describing an animal, is to be sure and give its character and spirit, for in that you have, without error, the sum and effect of all its parts, known and unknown. You must tell what it is to man. Surely the most important part of an animal is its *anima*, its vital spirit, on which is based its character and all the peculiarities by which it most concerns us. Yet most scientific books which treat of animals leave this out altogether, and what they describe are as it were phenomena of dead matter. What is most interesting in a dog, for example, is his attachment to his master, his intelligence, courage, and the like, and not his anatomical structure or even many habits which affect us less.

If you have undertaken to write the biography of an animal, you will have to present to us the living creature, *i.e.*, a result which no man can understand, but only in his degree report the impression made on him.

Science in many departments of natural history does not pretend to go beyond the shell; *i.e.*, it does not get to animated nature at all. A history of animated nature must itself be animated.

The ancients, one would say, with their gorgons, sphinxes, satyrs, mantichora, etc., could imagine more than existed, while the moderns cannot imagine so much as exists.

18 February 1860, *Journal* XIII:154–55

A fact stated barely is dry. It must be the vehicle of some humanity in order to interest us. It is like giving a man a stone when he asks you for bread. Ultimately the moral is all in all, and we do not mind it if inferior truth is sacrificed to superior, as when the moralist fables and makes animals speak and act like men. It must be warm, moist, incarnated, —have been breathed on at least. A man has not seen a thing who has not felt it.

23 February 1860, *Journal* XIII:160

As it is important to consider Nature from the point of view of science, remembering the nomenclature and system of men, and so, if possible, go a step further in that direction, so it is equally important often to ignore or forget all that men presume that they know, and take an original and unprejudiced view of Nature, letting her make what impression she will on you, as the first men, and all children and natural men still do. For our science, so called, is always more barren and mixed up with error than our sympathies are.

28 February 1860, *Journal* XIII:168–69

The old naturalists were so sensitive and sympathetic to nature that they could be surprised by the ordinary events of life. It was an incessant miracle to them, and therefore gorgons and flying dragons were not incredible to them. The greatest and saddest defect is not credulity, but our habitual forgetfulness that our science is ignorance. . . .

So far as the natural history is concerned, you often have your choice between uninteresting truth and interesting falsehood.

5 March 1860, *Journal* XIII:180–81

See at Lee's a pewee (phoebe) building. She has just woven in, or laid on the edge, a fresh sprig of saxifrage in flower. I notice that phoebes will build in the same recess in a cliff year after year. It is a constant thing here, though they are often disturbed. Think how many pewees must have built under the eaves of this cliff since pewees were created and this cliff itself built!! You can possibly find the crumbling relics of how many, if you should look carefully enough! It takes us many years to find out that Nature repeats herself annually. But how perfectly regular and calculable all her phenomena must appear to a mind that has observed her for a thousand years!

5 May 1860, *Journal* XIII:278–79

"Let me lead you back into your wood-lots again"

In my capacity of surveyor, I have often talked with some of you, my employers, at your dinner-tables, after having gone round and round and behind your farming, and ascertained exactly what its limits were. Moreover, taking a surveyor's and a naturalist's liberty, I have been in the habit of going across your lots much oftener than is usual, as many of you, perhaps to your sorrow, are aware. Yet many of you, to my relief, have seemed not to be aware of it; and, when I came across you in some out-of-the way nook of your farms, have inquired, with an air of surprise, if I were not lost, since you had never seen me in that part of the town or county before; when, if the truth were known, and it had not been for betraying my secret, I might with more propriety have inquired if *you* were not lost, since I had never seen *you* there before. I have several times shown the proprietor the shortest way out of his wood-lot.

Therefore, it would seem that I have some title to speak to you to-day; and considering what that title is, and the occasion that has called us together, I need offer no apology if I invite your attention, for the few moments that are allotted me, to a purely scientific subject.

At those dinner-tables referred to, I have often been asked, as many of you have been, if I could tell how it happened, that when a pine wood was cut down an oak one commonly sprang up, and *vice versa*. To which I have answered, and now answer, that I can tell, — that it is no mystery to me. As I am not aware that this has been clearly shown by any one, I shall lay the more stress on this point. Let me lead you back into your wood-lots again.

When, hereabouts, a single forest tree or a forest springs up naturally where none of its kind grew before, I do not hesitate to say, though in some quarters still it may sound paradoxical, that it came from a seed. Of the various ways by which trees are *known* to be propagated, — by transplanting, cuttings, and the like, — this is the only supposable one under these circumstances. No such tree has ever been known to spring from anything else. If any one asserts that it sprang from something else, or from nothing, the burden of proof lies with him.

It remains, then, only to show how the seed is transported from where it grows to where it is planted. This is done chiefly by the agency of the wind, water, and animals. The lighter seeds, as those of pines and maples, are transported chiefly by wind and water; the heavier, as acorns and nuts, by animals.

In all the pines, a very thin membrane, in appearance much like an insect's wing, grows over and around the seed, and independent of it, while the latter is being developed within its base. Indeed this is often perfectly developed, though the seed is abortive; nature being, you would say, more sure to provide the means of transporting the seed, than to provide the seed to be transported. In other words, a beautiful thin sack is woven around the seed, with a handle to it such as the wind can take hold of, and it is then committed to the wind, expressly that it may transport the seed and extend the range of the species; and this it does, as effectually as when seeds are sent by mail in a different kind of sack from the patent-office. There is a patent-office at the seat of government of the universe, whose managers are as much interested in the dispersion of seeds as anybody at Washington can be, and their operations are infinitely more extensive and regular.

There is then no necessity for supposing that the pines have sprung up from nothing, and I am aware that I am not at all peculiar in asserting that they come from seeds, though the mode of their propagation *by nature* has been but little attended to. They are very extensively raised from the seed in Europe, and are beginning to be here.

When you cut down an oak wood, a pine wood will not *at once* spring up there unless there are, or have been quite recently, seed-bearing pines near enough for the seeds to be blown from them. But, adjacent to a forest of pines, if you prevent other crops from growing there, you will surely have an extension of your pine forest, provided the soil is suitable.

As for the heavy seeds and nuts which are not furnished with wings, the notion is still a very common one that, when the trees which bear these spring up where none of their kind were noticed before, they have come from seeds or other principles spontaneously generated there in an unusual manner, or which have lain dormant in the soil for centuries, or perhaps been called into activity by the heat of a burning. I do not believe these assertions, and I will state some of the ways in which, according to my observation, such forests are planted and raised.

Every one of these seeds, too, will be found to be winged or legged in another fashion. Surely it is not wonderful that cherry-trees of all kinds are widely dispersed, since their fruit is well known to be the favorite food of various birds. Many kinds are called bird-cherries, and they appropriate many more kinds, which are not so called. Eating cherries is a bird-like employment, and unless we disperse the seeds occasionally, as they do, I shall think that the birds have the best right to them. See how artfully the seed of a cherry is placed in order that a bird may be compelled to transport it — in the very midst of a tempting pericarp, so that the creature that would devour this must commonly take the stone also into its mouth or bill. If you ever ate a cherry, and did not make two bites of it, you must have perceived it—right in the centre of the luscious morsel, a large earthy residuum left on the tongue. We thus take into our mouths cherry stones as big as peas, a dozen at once, for Nature can persuade us to do almost anything when she would compass her ends. Some wild men and children instinctively swallow these, as the birds do when in a hurry, it being the shortest way to get rid of them. Thus, though these seeds are not provided with vegetable wings, Nature has impelled the thrush tribe to take them into their bills and fly away with them; and they are winged in another sense, and more effectually than the seeds of pines, for these are carried even against the wind. The consequence is, that cherry-trees grow not only here but there. The same is true of a great many other seeds.

But to come to the observation which suggested these remarks. As I have said, I suspect that I can throw some light on the fact, that when hereabouts a dense pine wood is cut down, oaks and other hard woods may at once take its place. I have got only to show that the acorns and nuts, provided they are grown in the neighborhood, are regularly planted in such woods; for I assert that if an oak-tree has not grown within ten miles, and man has not carried acorns thither, then an oak wood will not spring up *at once*, when a pine wood is cut down.

Apparently, there were only pines there before. They are cut off, and after a year or two you see oaks and other hard woods springing up there, with scarcely a pine amid them, and the wonder commonly is, how the seed could have lain in the ground so long without decaying. But the

truth is, that it has not lain in the ground so long, but is regularly planted each year by various quadrupeds and birds.

In this neighborhood, where oaks and pines are about equally dispersed, if you look through the thickest pine wood, even the seemingly unmixed pitch-pine ones, you will commonly detect many little oaks, birches, and other hard woods, sprung from seeds carried into the thicket by squirrels and other animals, and also blown thither, but which are overshadowed and choked by the pines. The denser the evergreen wood, the more likely it is to be well planted with these seeds, because the planters incline to resort with their forage to the closest covert. They also carry it into birch and other woods. This planting is carried on annually, and the oldest seedlings annually die; but when the pines are cleared off, the oaks, having got just the start they want, and now secured favorable conditions, immediately spring up to trees.

The shade of a dense pine wood is more unfavorable to the springing up of pines of the same species than of oaks within it, though the former may come up abundantly when the pines are cut, if there chance to be sound seed in the ground.

But when you cut off a lot of hard wood, very often the little pines mixed with it have a similar start, for the squirrels have carried off the nuts to the pines, and not to the more open wood, and they commonly make pretty clean work of it; and moreover, if the wood was old, the sprouts will be feeble or entirely fail; to say nothing about the soil being, in a measure, exhausted for this kind of crop.

If a pine wood is surrounded by a white oak one chiefly, white oaks may be expected to succeed when the pines are cut. If it is surrounded instead by an edging of shrub-oaks, then you will probably have a dense shrub-oak thicket.

I have no time to go into details, but will say, in a word, that while the wind is conveying the seeds of pines into hard woods and open lands, the squirrels and other animals are conveying the seeds of oaks and walnuts into the pine woods, and thus a rotation of crops is kept up.

❦

Though I do not believe that a plant will spring up where no seed has been, I have great faith in a seed — a, to me, equally mysterious origin for it. Convince me that you have a seed there, and I am prepared to expect wonders. I shall even believe that the millennium is at hand, and that the reign of justice is about to commence, when the Patent Office, or Government, begins to distribute, and the people to plant, the seeds of these things.

In the spring of 1857 I planted six seeds sent to me from the Patent Office, and labeled, I think, "*Poitrine jaune grosse*," large yellow squash. Two came up, and one bore a squash which weighed 123½ pounds, the other bore four, weighing together 186¼ pounds. Who would have believed that there was 310 pounds of *poitrine jaune grosse* in that corner of my garden? These seeds were the bait I used to catch it, my ferrets which I sent into its burrow, my brace of terriers which unearthed it. A little mysterious hoeing and manuring was all the *abracadabra presto-change* that I used, and lo! true to the label, they found for me 310 pounds of *poitrine jaune grosse* there, where it never was known to be, nor was before. These talismans had perchance sprung from America at first, and returned to it with unabated force. The big squash took a premium at your fair that fall, and I understood that the man who bought it, intended to sell the seeds for ten cents a piece. (Were they not cheap at that?) But I have more hounds of the same breed. I learn that one which I despatched to a distant town, true to its instincts, points to the large yellow squash there, too, where no hound ever found it before, as its ancestors did here and in France.

Other seeds I have which will find other things in that corner of my garden, in like fashion, almost any fruit you wish, every year for ages, until the crop more than fills the whole garden. You have but little more to do than throw up your cap for entertainment these American days. Perfect alchemists I keep who can transmute substances without end, and thus the corner of my garden is an inexhaustible treasure-chest. Here you can dig, not gold, but the value which gold merely represents; and there is no Signor Blitz about it. Yet farmers' sons will stare by the hour to see a juggler draw ribbons from his throat, though he tells them it is all deception. Surely, men love darkness rather than light.

"The Succession of Forest Trees," 20 September 1860
(*Excursions*, 226–33, 248–50)

In August, '55, I levelled for the artificial pond at Sleepy
Hollow. They dug gradually for three or four years and
completed the pond last year, '59. It is now about a dozen
rods long by five or six wide and two or three deep, and
is supplied by copious springs in the meadow. There is a
long ditch leading into it, in which no water now flows,
nor has since winter at least, and a short ditch leading out
of it into the brook. It is about sixty rods from the very
source of the brook. Well, in this pond thus dug in the
midst of a meadow a year or two ago and supplied by
springs in the meadow, I find to-day several small patches
of the large yellow and the kalmiana lily already estab-
lished. Thus in the midst of death we are in life. The water is otherwise apparently clear of weeds. The river,
where these abound, is about half a mile distant down the
little brook near which this pond lies, though there *may*
be a few pads in the ditched part of it at half that distance.
How, then, did the seed get here? I learned last winter
(*vide* December 23, 1859) that many small pouts and
some sizable pickerel had been caught here, though the
connection with the brook is a very slight and shallow
ditch. I think, therefore, that the lily seeds have been con-
veyed into this pond from the river immediately, or perchance from the meadow between, either by fishes, reptiles, or birds which fed on them, and that the seeds were
not lying dormant in the mud. You have only to dig a
pond anywhere in the fields hereabouts, and you will
soon have not only water-fowl, reptiles, and fishes in it,
but also the usual water-plants, as lilies, etc. You will no
sooner have got your pond dug than nature will begin to
stock it. I suspect that turtles eat these seeds, for I often
see them eating the decayed lily leaves. If there is any wa-

ter communication, perhaps fishes arrive first, and then the water-plants for their food and shelter.

10 October 1860, *Journal* XIV:109–10

The scientific differs from the poetic or lively description somewhat as the photographs, which we so weary of viewing, from paintings and sketches, though this comparison is too favorable to science. All science is only a makeshift, a means to an end which is never attained. After all, the truest description, and that by which another living man can most readily recognize a flower, is the unmeasured and eloquent one which the sight of it inspires. No scientific description will supply the want of this, though you should count and measure and analyze every atom that seems to compose it.

Surely poetry and eloquence are a more universal language than that Latin which is confessedly dead. In science, I should say, all description is postponed till we know the whole, but then science itself will be cast aside. But unconsidered expressions of our delight which any natural object draws from us are something complete and final in themselves, since all nature is to be regarded as it concerns man; and who knows how near to absolute truth such unconscious affirmations may come? Which are the truest, the sublime conceptions of Hebrew poets and *seers*, or the guarded statements of modern geologists, which we must modify or unlearn so fast?

As they who were present early at the discovery of gold in California, and observed the sudden fall in its value, have most truly described that state of things, so it is commonly the old naturalists who first received American plants that describe them best. A scientific description is such as you would get if you should send out the scholars of the polytechnic school with all sorts of metres made and patented to take the measures for you of any natural object. In a sense you have got nothing new thus, for every object that we see mechanically is mechanically daguerreotyped on our eyes, but a true description growing out [of] the perception and appreciation of it is itself a new fact, never to be daguerreotyped, indicating the highest quality of the plant, — its relation to man, — of far more importance than any merely medicinal quality that it may possess, or be thought to-day to possess. There is a certainty and permanence about this kind of observation, too, that does not belong to the other, for every flower and weed has its day in the medical pharmacopoeia, but the beauty of flowers is perennial in the taste of men. . . .

The one description interests those chiefly who have not seen the thing; the other chiefly interests those who have seen it and are most familiar with it, and brings it home to the reader. We like to read a good description of no thing so well as of that which we already know the best, as our friend, or ourselves even. In proportion as we get and are near to our object, we do without the measured or scientific account, which is like the measure they take, or the description they write, of a man when he leaves his country, and insert in his passport for the use of the detective police of other countries. The men of science merely look at the object with sinister eye, to see if [it] corresponds with the passport, and merely visé or make some trifling additional mark on its passport and let it go; but the real acquaintances and friends which it may have in foreign parts do not ask to see nor think of its passport.

Gerard has not only heard of and seen and raised a plant, but felt and smelled and tasted it, applying all his senses to it. You are not distracted from the thing to the system or arrangement. In the true natural order the order or system is not insisted on. Each is first, and each last. That which presents itself to us this moment occupies the whole of the present and rests on the very topmost point of the sphere, under the zenith. The species and individuals of all the natural kingdoms ask our attention and admiration in a round robin. We make straight lines, putting a captain at their head and a lieutenant at their tails, with sergeants and corporals all along the line and a flourish of trumpets near the beginning, insisting on a particular uniformity where Nature has made curves to which belongs their own sphere-music. It is indispensable for us to square her circles, and we offer our rewards to him who will do it.

Who [*sic*] describes the most familiar object with a zest and vividness of imagery as if he saw it for the first time, the novelty consisting not in the strangeness of the object, but in the new and clearer perception of it.

13 October 1860, *Journal* XIV, 117–20

Our wood-lots, of course, have a history, and we may often recover it for a hundred years back, though we *do* not. A small pine lot may be a side of such an oval, or a half, or a square in the inside with all the curving sides cut off by fences. Yet if we attended more to the history of our lots we should manage them more wisely. . . .

I have come up here this afternoon to see ———'s dense white pine lot beyond the pond, that was cut off last winter, to know how the little oaks look in it. To my surprise and chagrin, I find that the fellow who calls him-

self its owner has burned it all over and sowed winter-rye
here. He, no doubt, means to let it grow up again in a year
or two, but he thought it would be clear gain if he could
extract a little rye from it in the meanwhile. What a fool!
Here nature had got everything ready for this emergency,
and kept them ready for many years, — oaks half a dozen
years old or more, with fusiform roots full charged and
tops already pointing skyward, only waiting to be
touched off by the sun, — and he thought he knew better,
and would get a little rye out of it first, which he could
feel at once between his fingers, and so he burned it, and
dragged his harrow over it. As if oaks would bide *his* time
or come at his bidding. Or as if he preferred to have a
pine or a birch wood here possibly half a century hence
— for the land is "pine sick" — rather than an oak wood
at once. So he trifles with nature. I am chagrined for him.
That he should call himself an agriculturalist! He needs
to have a guardian placed over him. A forest-warden
should be appointed by the town. Overseers of poor
husbandmen.

He has got his dollars for the pine timber, and now he
wishes to get his bushels of grain and finger the dollars
that they will bring; and then, Nature, you may have your
way again. Let us purchase a mass for his soul. A greedi-
ness that defeats its own ends. . . .

As time elapses, and the resources from which our
forests have been supplied fail, we shall of necessity be
more and more convinced of the significance of the seed.

16 October 1860, *Journal* XIV:125–26, 130–32

I see spatter-dock pads and pontederia in that little pool
at south end of Beck Stow's. How did they get there?

There is no stream in this case? It was perhaps rather reptiles and birds than fishes, then. Indeed we might as well ask how they got anywhere, for all the pools and fields have been stocked thus, and we are not to suppose as many new creations as pools. This suggests to inquire how any plant came where it is,—how, for instance, the pools which were stocked with lilies before we were born or this town was settled, and ages ago, were so stocked, as well as those which we dug. I think that we are warranted only in supposing that the former was stocked in the same way as the latter, and that there was not a sudden new creation,—at least since the first; yet I have no doubt that peculiarities more or less considerable have thus been gradually produced in the lilies thus planted in various pools, in consequence of their various conditions, though they all came originally from one seed.

We find ourselves in a world that is already planted, but is also still being planted as at first. We say of some plants that they grow in wet places and of others that they grow in desert places. The truth is that their seeds are scattered almost everywhere, but here only do they succeed. Unless you can show me the pool where the lily was created, I shall believe that the oldest fossil lilies which the geologist has detected (if this is found fossil) originated in that locality in a similar manner to these of Beck Stow's. We see thus how the fossil lilies which the geologist has detected are dispersed, as well as these which we carry in our hands to church.

The development theory implies a greater vital force in nature, because it is more flexible and accommodating, and equivalent to a sort of constant *new* creation.

18 October 1860, *Journal* XIV:146–47

I examine that oak lot of Rice's next to the pine strip of the 16th. The oaks (at the southern end) are about a dozen years old. As I expected, I find the stumps of the pines which stood there before quite fresh and distinct, not much decayed, and I find by their rings that they were about forty years old when cut, while the pines which spring from [them] are now about twenty-five or thirty. But further, and unexpectedly, I find the stumps, in great numbers, now much decayed, of an oak wood which stood there more than sixty years ago. They are mostly shells, the sap-wood rotted off and the inside turned to mould. Thus I distinguished four successions of trees.

Thus I can easily find in countless numbers in our forest, frequently in the third succession, the stumps of the oaks which were cut near the end of the last century. Perhaps I can recover thus generally the oak woods of the beginning of the last century, if the land has remained woodland. I have an advantage over the geologist, for I can not only detect the order of events but the time dur-

ing which they elapsed, by counting the rings on the stumps. Thus you can unroll the rotten papyrus on which the history of the Concord forest is written.

19 October 1860, *Journal* XIV:151–52

P.M. — To Blood's oak lot. . . .

I was struck by the orderly arrangement of the trees, as if each knew its own place; and it was just so at Wetherbee's lot. This being an oak wood, and like that, somewhat meadow [*sic*] in the midst, the swamp white oaks with a very few maples occupied that part, and I think it likely that a similar selection of the ground might have been detected often in the case of the other oaks, as the white compared with the red. As if in the natural state of things, when sufficient time is given, trees will be found occupying the places most suitable to each, but when they are interfered with, some are prompted to grow where they do not belong and a certain degree of confusion is produced. That is, our forest generally is in a transition state to a settled and normal condition.

5 November 1860, *Journal* XIV:215, 218

How is any scientific discovery made? Why, the discoverer takes it into his head first. He must all but see it.

25 November 1860, *Journal* XIV:267

It is the discovery of science that stupendous changes in the earth's surface, such as are referred to the Deluge, for instance, are the result of causes still in operation, which have been at work for an incalculable period. There has not been a sudden re-formation, or, as it were, new creation of the world, but a steady progress according to existing laws. The same is true in detail also. It is a vulgar prejudice that some plants are "spontaneously generated," but science knows that they come from seeds, *i.e.* are the result of causes still in operation, however slow and unobserved. It is a common saying that "little strokes fall great oaks," and it does not imply much wisdom in him who originated it. The sound of the axe invites our attention to such a catastrophe; we can easily count each stroke as it is given, and all the neighborhood is informed by a loud crash when the deed is consummated. But such, too, is the rise of the oak; little strokes of a different kind and often repeated raise great oaks, but scarcely a traveller hears these or turns aside to converse with Nature, who is dealing them the while.

Nature is slow but sure; she works no faster than need be; she is the tortoise that wins the race by her perseverance; she knows that seeds have many other uses than to reproduce their kind. In raising oaks and pines, she works with a leisureliness and security answering to the age and strength of the trees. If every acorn of this year's crop is destroyed, never fear! she has more years to come. It is not necessary that a pine or an oak should bear fruit every year, as it is that a pea-vine should. So, botanically, the greatest changes in the landscape are produced more gradually than we expected. If Nature has a pine or an oak wood to produce, she manifests no haste about it.

❦

Thus we should say that oak forests are produced by a kind of accident, *i.e.* by the failure of animals to reap the fruit of their labors. Yet who shall say that they have not a fair knowledge of the value of their labors — that the squirrel when it plants an acorn, or the jay when it lets

one slip from under its foot, has not a transient thought
for its posterity?

14 January 1861, *Journal* XIV:311–13

Botanists talk about the possibility and impossibility of
plants being naturalized here or there. But what plants
have not been naturalized? Of course only those which
grow to-day exactly where the original plant of the
species was created. It is true we do not know whether
one or many plants of a given kind were originally created, but I think it is the most reasonable and simple to
suppose that only one was, — to suppose as little depar-
ture as possible from the existing order of things. They
commenced to spread themselves at once and by whatever means they possessed as far as they could, and they
are still doing so. Many were common to Europe and
America at the period of the discovery of the latter country, and I have no doubt that they had naturalized themselves in one or the other country. This is more philosophical than to suppose that there were independently
created in each. . . .

It is evident that Nature's designs would not be accomplished if seeds, having been matured, were simply
dropped and so planted directly beneath their parent
stems, as many will always be in any case. The next con-
sideration with her, then, after determining to create a
seed, must have been how to get it transported, though to
never so little distance, — the width of the plant, or less,
will often be sufficient, — even as the eagle drives her
young at last from the neighborhood of her eyrie, — for
their own good, since there is not food enough there for
all, — without depending on botanists, patent offices, and

seedsmen. It is not enough to have matured a seed which will reproduce its kind under favorable conditions, but she must also secure it those favorable conditions. Nature has left nothing to the mercy of man. She has taken care that a sufficient number of every kind of seeds, from a cocoanut to those which are invisible, shall be transported and planted in a suitable place.

A seed, which is a plant or tree in embryo, which has the principle of growth, of life, in it, is more important in my eyes, and in the economy of Nature, than the diamond of Kohinoor.

22 March 1861, Journal XIV:332–34

❦

"All this is perfectly distinct to an observant eye"

After a violent easterly storm in the night, which clears up at noon (November 3, 1861), I notice that the surface of the railroad causeway, composed of gravel, is singularly marked, as if stratified like some slate rocks, on their edges, so that I can tell within a small fraction of a degree from what quarter the rain came. These lines, as it were of stratification, are perfectly parallel, and straight as a ruler, diagonally across the flat surface of the causeway for its whole length. Behind each little pebble, as a protecting boulder, an eighth or a tenth of an inch in diameter, extends northwest a ridge of sand an inch or more, which it has protected from being washed away, while the heavy drops driven almost horizontally have washed out a furrow on each side, and on all sides are these ridges, half an inch apart and perfectly parallel.

All this is perfectly distinct to an observant eye, and

yet could easily pass unnoticed by most. Thus each wind is self-registering.

Final entry, 3 November 1861, *Journal* XIV:346

FURTHER READING

Thoreau, Henry David. *The Correspondence of Henry David Thoreau.* Edited by Walter Harding and Carl Bode. New York: New York University Press, 1958.

——. *Excursions.* Volume IX of *The Writings of Henry David Thoreau.* Boston: Houghton Mifflin, 1893.

——. *Journal.* 5 volumes to date. Various editors. Princeton: Princeton University Press, 1981–.

——. *The Journal of Henry D. Thoreau.* 14 volumes. Edited by Bradford Torrey and Francis Allen. Boston: Houghton Mifflin, 1906. Reprint, New York: Dover, 1962.

——. *Walden.* Edited by J. Lyndon Shanley. Princeton: Princeton University Press, 1971.

——. *A Week on the Concord and Merrimack Rivers.* Edited by Carl Hovde, William L. Howarth, and Elizabeth Hall Witherell. Princeton: Princeton University Press, 1980.

The Spirit of Thoreau

"How many a man has dated a new era in his life from the reading of a book," wrote Henry David Thoreau in *Walden*. Today that book, perhaps more than any other American work, continues to provoke, inspire, and change lives all over the world, and each rereading is fresh and challenging. Yet as Thoreau's countless admirers know, there is more to the man than *Walden*. An engineer, poet, teacher, naturalist, lecturer, and political activist, he truly had several lives to lead, and each one speaks forcefully to us today.

The Spirit of Thoreau introduces the thoughts of a great writer on a variety of important topics, some that we readily associate him with, some that may be surprising. Each book includes selections from his familiar published works as well as from less well known and even previously unpublished lectures, letters, and journal entries. Thoreau claimed that "to read well, that is, to read true books in a true spirit, is a noble exercise, and one that will task the reader more than any exercise which the customs of the day esteem." The volume editors and the Thoreau Society believe that you will find these new aspects of Thoreau an exciting "exercise" indeed.

This Thoreau Society series reunites Henry Thoreau with

his historic publisher. For more than a hundred years, the venerable publishing firm of Houghton Mifflin has been associated with standard editions of the works of Emerson and Thoreau and with important bibliographical and interpretive studies of the New England Transcendentalists. Until Princeton University Press began issuing new critical texts in *The Writings of Henry D. Thoreau,* beginning with *Walden* in 1971, Thoreauvians were well served by Houghton Mifflin's twenty-volume Walden or Manuscript Edition of *The Writings of Henry David Thoreau* (1906). Having also published Walter Harding's annotated edition of *Walden* (1995), Houghton Mifflin is again in the forefront of Thoreau studies.

You are invited to continue exploring Thoreau by joining our society. For well over fifty years we have presented publications, annual gatherings, and other programs to further the appreciation of Thoreau's thought and writings. And now we have embarked on a bold new venture. In partnership with the Walden Woods Project, the Thoreau Society has formed the Thoreau Institute, a research and educational center housing the world's greatest collection of materials by and about Thoreau. In ways that the author of *Walden* could not have imagined, his message is still changing lives in a brand-new era.

For membership information, write to the Thoreau Society, 44 Baker Farm, Lincoln, MA 01773-3004, or call 781-259-4750. To learn more about the Thoreau Institute, write to the same address; call 781-259-4700; or visit the Web site:

www.walden.org.

WESLEY T. MOTT
Series Editor
The Thoreau Society